D1457221

GPS
DECLASSIFIED

GPS

DECLASSIFIED

From Smart Bombs to Smartphones

Richard D. Easton and Eric F. Frazier

Foreword by Rick W. Sturdevant

Potomac Books
An imprint of the University of Nebraska Press

Library of Congress Cataloging-in-Publication Data
Easton, Richard D.
GPS declassified: from smart bombs to smartphones / Richard
D. Easton and Eric F. Frazier; foreword by Rick W. Sturdevant.
pages cm
Includes bibliographical references and index.
ISBN 978-1-61234-408-9 (cloth: alk. paper)
1. Global Positioning System—History. I. Title.
G109.5.E37 2013
910.285—dc23   2013023507

Set in Lyon Text by Laura Wellington.
Designed by J. Vadnais.

Contents

Illustrations

Foreword

For one who has been writing a portion of the official history of Global Positioning System (GPS) operations since the 1980s, it is startling to realize that young people entering college in 2013 have never known a world without GPS. Even those who are aware of the system's transparent presence on their personal phone devices, in their daily business transactions, or amid their recreational activities undoubtedly take its benefits for granted. It would be surprising if more than a few could explain, at even a rudimentary level, how this amazing space-based positioning, navigation, and timing (PNT) system works. How many know GPS is the world's only global utility? Who stops to remember that its signals are available free of charge to anyone with a GPS receiver?

Here, at last, Richard Easton and Eric Frazier present in plain, simple language how a PNT system originally developed for military purposes—one that Air Force Space Command continues to operate and maintain—became essential for countless civil and commercial activities around the world. The authors deftly place the concept and development of GPS within two broader historical contexts: navigation and robotic spaceflight. Their description of the dissimilar problems that compelled visionaries in each of the military services to pursue a three-dimensional positioning and navigation system substantiates the adage, sometimes attributed to Plato, that necessity is the mother of invention.

On the way to fostering what emerged as GPS, however, ample participation occurred to justify multiple paternal claims. Eventually, different individuals garnered high-level recognition based on, and bolstering, their respective claims. One (Roger Easton) received the 2004 National Medal of Technology from President George W. Bush. The National Academy of Engineering awarded two others (Ivan Getting and Brad Parkinson) the 2003 Charles Stark Draper Prize. All three, in recognition of their seminal GPS roles, became inductees to the National Inventors Hall of Fame. In 2012 the National Space Club named the same trio among the "GPS Originator Team" that received the prestigious Dr. Robert H. Goddard Memorial Trophy. Numerous others also contributed, without fanfare or subsequent recognition, to the conceptualization and development of GPS.

Easton and Frazier explore the debatable parentage of GPS through sources previously ignored by or unavailable to other scholars. While not the definitive history of the origins of GPS and its place in the centuries-old panoply of navigational systems, their study certainly advances our knowledge of the "who, what, when, where, why, and how" behind this amazing technological accomplishment. This book represents a solid foundation upon which future scholars can build their research and writing about GPS, or what has become more broadly identified as global navigation satellite system (GNSS) technology.

Beyond the origin of GPS and how it works, these authors deliver an impressive survey of the historical evolution of GPS applications among military, civil, and commercial users, not to mention private individuals. Although certainly not all-inclusive, their astounding coverage of the many ways in which people rely on precise PNT from outer space boggles the mind. When the authors describe the system's vulnerability to interference, whether intentional or natural, the potentially devastating military, societal, economic, and political effects of GPS disruption take on sinister proportions.

Whenever historians venture into the future, based on their understanding of the past and their perception of the present, they generally fare no better than nonhistorians. All confront largely incomprehensible terrain. Nonetheless, Easton and Frazier dare to conclude their GPS study with an overview of possibilities. Indeed, advocates of vector analysis in history might perceive that these two authors discern probabilities based on chronological patterns or trends—scientific, technological, economic, political, military, and social. Still, they recognize that if history teaches us anything, it is to remain watchful for unexpected twists and unanticipated turns. That is precisely what keeps past, present, and future particularly interesting and occasionally controversial, as the following narrative demonstrates.

Rick W. Sturdevant, PhD
Deputy Director of History
HQ Air Force Space Command

Acknowledgments

The authors would like the thank the following people who shared information through interviews or correspondence, provided documents or images, read portions of the manuscript, and offered suggestions or otherwise aided the research and writing of this book:

Roy Anderson (now deceased), radio-navigation pioneer at General Electric, and his wife, Gladys; Jonathan Betts, senior curator of horology, Royal Observatory, Greenwich, UK; Veronique Bohbot, PhD, professor of psychiatry at McGill University and Douglas Institute in Montreal; Walter Boyne, author/historian; Michael Buckley, public information officer, the Johns Hopkins University Applied Physics Laboratory; Alan Cameron, publisher and editor-in-chief, *GPS World*; Glen Gibbons, editor, *Inside GNSS*; David Gosch, senior public relations specialist, Rockwell Collins; Mike Gruntman, professor of astronautics, University of Southern California; R. Cargill Hall, historian emeritus, National Reconnaissance Office; Robert Kern, president, Kernco Inc., and builder of atomic frequency standards for the GPS program; Jason Kim, senior advisor, National Coordination Office for Space-Based PNT; Chester Kleczek (now deceased), former engineer at Naval Air Systems Command, who was the original program manager and sponsor of Timation; Arthur McCoubrey, cofounder of Frequency and Time Systems, which built early atomic frequency standards for the GPS program; Keith D. McDonald, former scientific director of the Navigation Satellite Executive Steering Group and executive secretary of the Defense Navigation Planning Group; Kimberly Morgan, corporate communications, Texas Instruments; Dian Moulin, daughter of Navy captain David Holmes (deceased); Harold Rosen, former vice president at Hughes Aircraft; Dava Sobel, author; Harry Sonnemann, former chairman of the Navigation Satellite Executive Steering Group; Rick Sturdevant, deputy command historian, U.S. Air Force Space Command, Peterson AFB, Colorado Springs, Colorado; Martin Votaw, former engineer at the Naval Research Laboratory; Phil Ward, president, Navward GPS Consulting; and current or former Naval Research Laboratory staffers Jonna Atkinson, Jamie Baker, Ron Beard, James Buisson, Dean Bundy, Gayle Fullerton, Lee Hammarstrom, Vijay Koweth, Thomas McCaskill, Kathy Parrish, Leo Slater, Richard Thompson, Jim Tugman, Joe White, Robert Whitlock, and Peter Wilhelm.

Additionally, Eric Frazier would like to thank his daughter, Carolyn Frazier, for sparking his initial interest in writing a book about GPS and his son, Will Frazier, for providing research assistance. Richard Easton would like to thank his father, Roger Easton, and his sister, Ruth Easton, for their assistance in locating pertinent documents.

Abbreviations

2SOPS: 2nd Space Operations Squadron
ABMA: U.S. Army Ballistic Missile Agency
ALCM: air-launched cruise missile
APL: Applied Physics Laboratory
ARPA: Advanced Research Projects Agency
ASAT: antisatellite
AVL: automatic vehicle location
BCCI: Bank of Credit and Commerce International
CALCM: conventional air-launched cruise missile
CBO: Congressional Budget Office
CDMA: code division multiple access
CPU: central processing unit
CSEL: Combat Survivor Evader Locator
CTIA: Cellular Telecommunications Industry Association
DARPA: Defense Advanced Research Projects Agency
DART: Demonstration for Autonomous Rendezvous Technology
DGPS: differential GPS
DNSDP: Defense Navigation Satellite Development Plan
DNSS: Defense Navigation Satellite System
DOD: Department of Defense

DSARC: Defense Systems Acquisition Review Council

DSP: Defense Support Program

EASCON: Electronics and Aerospace Systems Convention

EGNOS: European Geostationary Overlay System

EOSAT: Earth Observation Satellite Inc.

ESA: European Space Agency

EU: European Union

FAA: Federal Aviation Administration

FBCB2: Force XXI Battle Command Brigade and Below

FCC: Federal Communications Commission

FDMA: frequency division code modulation

FOC: Full Operational Capability

GAGAN: GPS-Aided Geo-Augmented Navigation

GAO: General Accounting/Accountability Office

GBAS: ground-based augmentation system

GDM: Generalized Development Model

GIS: geographic information systems

GNSS: global navigation satellite system

GPS: Global Positioning System

ICAO: International Civil Aviation Organization

ICBM: intercontinental ballistic missile

IEEE: Institute of Electrical and Electronics Engineers

IGEB: Interagency GPS Executive Board

iGPS or HIGPS: High Integrity GPS

IGY: International Geophysical Year

INS: inertial navigation systems

IOC: Initial Operational Capability

IONDS: Integrated Operational Nuclear (Detonation) Detection System

IRBM: intermediate-range ballistic missile

IRNSS: Indian Regional Navigation Satellite System

ISS: inertial surveying systems

JPO: joint program office

LBS: location-based service

MBOC: multiplex binary offset carrier

MSAS: Multifunctional Transport Satellite Augmentation System

NANU: Notice Advisory to Navstar Users

NAPA: National Association of Public Administration

NASA: National Aeronautics and Space Administration
NAVCEN: Navigation Center
NAVSEG: Navigation Satellite Executive Steering Group
NAVSMO: Navigation Satellite Management Office
NAVSPASUR: Naval Space Surveillance System
NDAA: National Defense Authorization Act
NDGPS: Nationwide DGPS
NES: Navigation Experimental Satellites
NNSS: Naval Navigation Satellite System
NORAD: North American Aerospace Defense Command
NRC: National Research Council
NRL: Naval Research Laboratory
NTIA: National Telecommunications and Information Administration
NTS: Navigation Technology Satellite
OCS: Operational Control System
OCX: Operational Control Segment
PDAS: personal digital assistants
PLGR: precision lightweight GPS receiver
PND: personal navigation device
PNT: positioning, navigation, and timing
PPC: Portable Professional Computer
PPS: Precise Positioning Service
PRN: pseudorandom noise
QZSS: Quasi-Zenith Satellite System
RAIM: Receiver Autonomous Integrity Monitoring
RAM: random access memory
READI: Real-Time Earthquake Analysis for Disaster
SAMSO: Space and Missile Systems Organization
SBSS: Space Based Space Surveillance
SDI: Strategic Defense Initiative
SLAM: standoff land attack missile
SLGR: small, lightweight GPS receiver ("Slugger")
SLVS: space launch vehicles
SPS: Standard Positioning Service
SVN: space vehicle number
SVS: space vehicles
TERCOM: terrain contour matching

TLAM: Tomahawk land attack missile
UAV: unmanned aerial vehicle
UTC: Coordinated Universal Time
V2I, V2X: vehicle-to-infrastructure
V2V: vehicle-to-vehicle
VMT: vehicle-miles-traveled
WAAS: Wide Area Augmentation System
WGS-84: World Geodetic System 1984

GPS
DECLASSIFIED

Introduction

There is no "middle of nowhere" anymore, Agent Scully.

The Shadow Man in "Trust No. 1," X-Files, season 9, originally broadcast January 6, 2002

The stolen 2009 Chevrolet Tahoe was pushing 100 mph on a residential street in Visalia, California, by the time Sgt. Randy Lentzner and his partner, Officer Robert Gilson, got their police cruiser into position behind it.[1] High-speed chases often end in collisions and always endanger innocent bystanders, so the time—about 3:30 a.m.—was in their favor. But on this morning, October 18, 2009, the police had another advantage. Some 12,500 miles overhead, a network of satellites was helping to track the vehicle's exact position, and new technology was in place to eliminate the need for hazardous spike strips or bumping tactics traditionally used in police pursuits.

Minutes after a shotgun-wielding assailant forced the Tahoe's owner, Jose Ruiz, and his cousin out of the vehicle, Ruiz reported the theft to police and told them it was equipped with General Motors' OnStar navigation, communication, and security system. A call to OnStar quickly provided police with the location of the vehicle, and when the officers confirmed they were ready, OnStar call-center advisors transmitted a wireless signal to the Tahoe, activating its "Stolen Vehicle Slowdown" feature.

Behind the wheel of the Tahoe, the driver pressed the accelerator harder. To his surprise, the pedal became unresponsive, and the SUV's engine gradually slowed to idle speed. He jumped out and ran but was soon in handcuffs—roughly fifteen minutes after Ruiz reported his vehicle stolen. Twenty-one-year-old Albert Roman Romero became the first person caught and charged with carjacking as a direct result of GPS technology.[2]

The Global Positioning System, now so universally recognized that the Associated Press uses the GPS acronym without elaboration, was not yet complete when Romero was born.[3] The first GPS satellite was launched in 1978, two decades after the Soviet Union surprised the West with its launch of Sputnik, but the twenty-fourth satellite—the minimum number required to provide uninterrupted, round-the-clock coverage worldwide—did not go into orbit until 1994. That was three years after the Soviet Union dissolved and three decades after the first conceptual schemes for such a system emerged in the 1960s.

Over the course of five decades, the development of GPS has featured scientific genius and foresight, interservice rivalries among the branches of the military, defense contractor controversies, cancellations, delays and budget cuts, stunning success on the battlefield, and impressive entrepreneurial innovation—all against the backdrop of shifting foreign policy, wars, recessions, and unexpected events beyond the confines of the program. By the fall of 2009, when OnStar recorded that first successful carjacking intervention, what began as a classified Cold War military program had spawned a private-sector GPS industry with a multitude of uses dwarfing those related to defense and a global market in the tens of billions of dollars. Despite varying estimates of the size of the global market, most analysts predict a combined annual growth rate above 20 percent through 2016.[4]

GPS is at once a simple concept and a vastly complex technology. Stripped of the intricate math and physics—in this case, it *is* rocket science that makes the system possible—GPS may best be understood as a set of free radio signals available worldwide that enables scores of individual applications requiring precise positioning, navigation, or timing. Although GPS achieved fame first for its ability to guide bombs accurately and later for changing the way people find their way around, its ability to synchronize time across vast distances—a key element of its origin—now enables the smooth flow of electronic data across worldwide networks. The degree to which private industry has leveraged this free service is phenomenal, making GPS a vital public utility today.

For example, OnStar, with more than 2 million subscribers in 2009, generated a billion dollars for GM, even as the automaker faced bankruptcy and survived only with a federal government bailout.[5] By 2013 OnStar boasted more than 6.4 million subscribers in the United States, Canada, and China, and nearly a million GM vehicle owners used its RemoteLink mobile phone app to remotely access OnStar services.[6] The company also created FMV (For My Vehicle), an aftermarket replacement rearview mirror with an OnStar button, making many of its services available in other manufacturers' vehicles, but rival carmakers have quickly ramped up their own similar systems. Considering that more than four hundred people die and many more sustain injuries each year in high-speed police pursuits, it is no surprise that law enforcement officials share vehicle owners' enthusiasm for the vehicle slowdown feature.[7] "It helped us not only recover a vehicle for a local citizen, but also prevented a dangerous high-speed chase and allowed us to quickly apprehend a suspect," said Sgt. Steven Phillips of the Visalia police. "It's a win for everyone."[8]

Or nearly everyone. Cable news channels televise high-speed police chases hoping viewers will stay glued to their screens in anticipation of a dramatic conclusion. The advent of electronic news gathering using helicopters made live aerial coverage of automobile chases a common feature of television newscasts by the mid-1990s.[9] With the emergence of a twenty-four-hour news cycle, the ability of local news operations to instantly uplink their video to national cable networks, and the celebrity factor embodied in the iconic 1994 slow-speed chase of O. J. Simpson's white Ford Bronco, cable news found a winning formula that presaged the era of "reality" TV. Cable executives may appreciate this irony: just as satellite technology enabled video crews to take live coverage aloft, the ever-expanding use of GPS satellites now offers the means to make high-speed police pursuits obsolete.[10]

A few naysayers worry that GPS will do to our innate sense of direction what keyboards have done to penmanship. Longstanding concerns persist about the technology's misuse by terrorists or governments, even against their own citizens. But for the most part, the march of GPS technology from the laboratory to the battlefield to everyday life has gained momentum with each passing year, especially since 2000, when the government stopped deliberately degrading the signal provided for civilian use. The boost in accuracy unleashed a torrent of consumer electronics aimed at helping motorists find their way, avoid traffic jams, and locate points of interest. Industry watchers estimated at the end of 2009 that more than a third of U.S. households had at least one personal navigation device (PND), and when in-dash vehicle systems and GPS-enabled phones were included, the figure rose to more than 55 percent.[11]

When CNN launched, in mid-2008, a new program covering world events, *Fareed Zakaria GPS*, viewers had no trouble making the connection when the host announced, "Welcome to the very first edition of 'Global Public Square.'"[12]

By the end of 2009, GPS had become such a household word—used interchangeably for the system and the gadgets that use it—that numerous television advertisers were tying their products to its popularity:

> Big-box jewelry retailer Jared showed a man sitting in his car asking a voice navigation system for directions. The unit's sultry female voice comments on his purchase and refuses to cooperate until he placates "her" by hanging a necklace around the GPS unit.[13]

> TurboTax touted its tax preparation software as being as easy to follow as GPS: "These days, if I need to get someplace, I just use the GPS on my cell phone. I get turn-by-turn directions, which show me right where I need to go. I do my taxes the same way, with the TurboTax Federal Free Edition."[14]

> Fidelity Investments adopted a moving green line and the slogan "Turn Here" to promote its financial and retirement planning services, a clear allusion to turn-by-turn GPS guidance.[15]

An *ABC News* online article in December 2009 listed GPS among its "Top 10 Innovations of the Decade."[16] A WashingtonPost.com blogger that same month asked his online readers to rank a list of the decade's top ten consumer tech developments. One week into the voting GPS placed fourth, behind the iPhone, Mozilla Firefox, and the iTunes store.[17]

For individuals the GPS revolution is most visible in how they shop and travel in their cars and on foot. From 2005 to 2010, consumers accounted for 59 percent of GPS equipment revenues in North America.[18] Mobile phone apps are increasing consumer exposure to GPS. By January 2013, 129.4 million Americans owned smartphones, or about 55 percent of the nation's 235 million mobile phone subscribers.[19] By 2016 there are projected to be 1 billion smartphones in use worldwide and 340 million mobile subscribers using a turn-by-turn navigation app or service.[20] While analysts expect PND sales to decline by about 40 percent through 2016 as users shift to phones, at the end of 2011 there were 150 million PNDs in use worldwide and 60 million factory-installed and aftermarket in-dash systems, ensuring they, too, will remain a key part of many people's daily lives.[21]

Most GPS users need not know and probably do not care how the technology works or where the satellites and ground stations transmitting the signals are located. But casual users should be aware that the satellite platform they rely upon for GPS service is one more example of public infrastructure—like bridges, highways, and water mains—that ages and that government finds increasingly costly and challenging to maintain. Those who do not personally own or use GPS devices—from the standpoint of mobile phones, that is becoming a rarity—might consider the issue irrelevant, but they should know that the technology now touches virtually everyone and every sector of society. In addition to military use and personal navigation, GPS has become indispensable for a host of commercial applications in aviation and space operations, trucking and shipping, fishing and boating, agriculture and forestry, surveying and mapping, grading and construction, and mining and oil exploration. Beyond positioning and navigation, GPS now provides atomic-clock accuracy for the synchronization and split-second timing needs of telecommunications and data systems, financial networks, and electric power grids.[22] A recent study estimated that GPS technology would provide U.S. commercial users annual direct economic benefits between $67.6 billion and $122.4 billion as its adoption by commercial users approaches 100 percent.[23]

As entrepreneurs have imagined new uses for GPS, demand for better accuracy has led to multiple ground- and space-based systems that augment the satellite signals. Other nations are developing their own versions of GPS, giving rise to the generic term *GNSS*, for any global navigation satellite system. While nations pledge to make their GNSS systems work together, they actively compete with one another as they do in all other areas. These systems contribute to an increasingly interconnected and complex world in which the systems' interoperability, sustainability, and security create political and economic issues affecting the entire population.

Recognizing the growing demands for positioning, navigation, and timing (PNT) services and the need to set clear policies involving the use of GPS, President George W. Bush issued a directive in 2004 creating a National Executive Committee for Space-Based PNT.[24] President Barack Obama reaffirmed the commitment to GPS in 2010 as part of his National Space Policy directive.[25] Addressing the PNT advisory board in November 2009, Scott Pace, director of the Space Policy Institute at George Washington University, said that global navigation satellite systems today are better understood as information technology infrastructure than merely as aerospace products.[26] With such a large installed user base, he said, introducing new signals or systems resembles rolling out new computer operating software, with similar concerns about backward compatibility and users' willingness and ability to upgrade. Everyone who has endured the learning curve of a new software program or deliberated over the purchase of a new computer that would necessitate costly upgrades can identify with that statement and appreciate where the GPS marketplace now stands. But, far more than the computer or consumer electronics industries, how GPS evolves depends on what the government does. That makes understanding the history of GPS development helpful for anyone hoping to construct informed opinions about public policies regarding its future.

This book attempts to present that history—from the scientific breakthroughs that made such a system possible to the people and institutions that oversaw its development—in an accessible manner for a general audience. As with planning a trip using GPS guidance, multiple options were available to tell this story. This work aims to strike a balance between following the shortest, most direct route and making sure readers visit the most important points of interest. At the end, you should have a better understanding of a technology that continues to revolutionize how humans travel, how we work and play, and perhaps even how we think.

New Moons Rising
The Satellite Age Arrives

From the vantage point of 2100 A.D., the year 1957 will most certainly stand in history as the year of man's progression from a two-dimensional to a three-dimensional geography. It may well stand also, as the point in time at which intellectual achievement forged ahead of weapons and national wealth as instruments of national policy.

Geophysicist Lloyd V. Berkner, Foreign Affairs, *January 1958*

Two seconds after liftoff and four feet into its flight, the Vanguard rocket stalled as if held by an invisible tether and sagged back toward the launch pad. As the bottom of the first stage, filled with liquid oxygen and kerosene propellant, crumpled against the platform, the engine's orange exhaust erupted into a massive fireball that engulfed the tottering seventy-two-foot rocket. In the final moment before flames obscured its tip, the nose cone could be seen breaking free.[1] Under that cone was a grapefruit-size aluminum sphere with batteries and transmitters packed inside and a half-dozen solar cells and antennas attached outside. The entire twenty-two-thousand-pound, three-stage rocket assembly, which an Associated Press story described as "a ton of metal and

10 tons of fuel," had been designed for the sole purpose of hurling the three-and-a-quarter-pound ball three hundred miles above Earth.[2] For the United States, the effort of putting its first satellite into orbit had gotten off to a dismal start—in full view of millions of Americans who watched televised images of the conflagration on the evening news on December 6, 1957.

Few people watching that night would have believed that, within a decade, the same scientist who designed that first tiny U.S. satellite would conceive a multi-satellite system for determining the precise location and exact time anywhere on the planet or in the air. By the time the forerunners of GPS were hitting the drawing boards, the public had shifted its attention to the race to put men on the moon. Plans for a global satellite-based navigation system, largely classified and visionary beyond what most of the military brass saw as practical, would have to wait years for technical advances and conventional thinking to catch up. However, any discussion of GPS must begin with the launching of the first man-made satellites at the dawn of the space age. The system's DNA traces directly back to the technologies, the scientists, and the military institutions that participated in what sometimes has been called "the first space race."[3]

After crews extinguished the fire and began cleaning up the Vanguard launch site, they found the satellite lying on the ground. It survived a seven-story fall, a 3,500-degree inferno, and being doused with tons of water. Its antennas were bent but the sphere was intact, and ground receivers set up to monitor its orbits confirmed that it was transmitting two radio signals, as designed.[4] Martin Votaw, who built the small transmitters (using early transistors in place of vacuum tubes), was listening as the satellite made its short journey to the ground. Votaw went to the launch pad to retrieve it. "There it was, clean as a whistle," he recalled clearly in an interview fifty-two years later.[5] He placed the battered satellite in a brown cardboard box and took it to the man who led its design team, thirty-six-year-old Roger Easton, future head of the Space Applications Branch at the Naval Research Laboratory (NRL).

"What should we do with it?" Votaw asked.

"Take it home, I guess," Easton replied.[6]

In a move that today would provoke intense questioning, if not an airport lockdown, Easton nonchalantly carried the box with the satellite aboard a commercial flight back to Washington DC.[7] "It sat on our kitchen table overnight," his daughter, Ruth, recalled.[8] Easton delivered the satellite to John P. Hagen, director of Project Vanguard, who later donated it to the Smithsonian's National Air and Space Museum, where it remains today.[9]

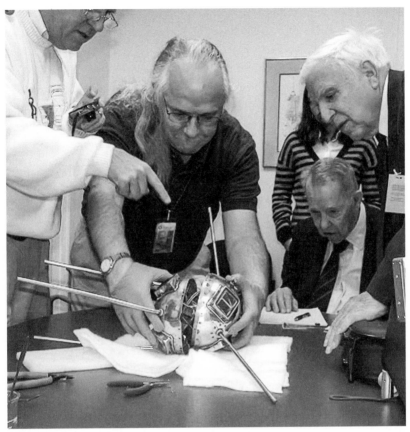

Fig. 1.1. Roger Easton (*upper right*) and Martin Votaw (*below, seated*) look on as the National Air and Space Museum's David Devorkin opens the Vanguard TV-3 satellite in 2008, during a fiftieth anniversary gathering. Chris Hagen, son of Vanguard Project director John P. Hagen, points at the artifact. (Courtesy Roger Easton Jr.)

Bad Press

While Vanguard personnel busied themselves repairing the launch facility and readying a backup rocket, the public reacted with panic. The incident plunged the nation, already in a state of high anxiety following the Soviet Union's two successful Sputnik launches, on October 4 and November 3, into a period of humiliation, second guessing, finger pointing, and political jockeying. "Vanguard Rocket Burns on Beach; Failure to Launch Test Satellite Assailed as Blow to U.S. Prestige," read the headline in the *New York Times* the next morning.[10] "Oh What a Flopnik!" chided the *London Daily Herald*.[11] "How about some relentless looking around for possible sabotage?" the *New York Daily News* asked.[12] The Baltimore-based Glenn L. Martin Company, the prime contractor for the Vanguard rockets, and General Electric, the subcontractor for the engines, blamed each other for the problem.[13] Sell orders forced New York Stock Exchange officials to halt trading in Martin's stock the day after the explosion.[14] The launch pad fiasco came just eleven days after a brash Texas senator with presidential ambitions, Lyndon Baines Johnson, began hearings on Sputnik in the Senate Preparedness Subcommittee. That same day, November 25, President Dwight D. Eisenhower suffered a mild stroke. Many wondered if he was healthy enough to continue as president, and there were some calls for him to step down.[15] American preeminence in world affairs and military and technological prowess, taken for granted since the end of World War II, was being openly questioned, as evidenced by Johnson remarking after the Vanguard failure, "How long, how long, oh God, how long will it take us to catch up with the Russians' two satellites?"[16]

From a modern perspective, with partisan bickering over foreign affairs routine and lack of faith in government pervasive, the reaction to the Sputnik launches and Vanguard failure may not seem unusual, but Americans viewed the federal government differently then. A 2013 Pew Research Center survey found only 26 percent of Americans trusted the government to do what is right "just about always" or "most of the time." That is slightly better than 2010, when the figure was 22 percent, but the trend has been mostly downward for decades. When Pew first polled the question of trust in government in 1958, as part of the National Election Study, 73 percent said they trusted the government to do what is right most of the time.[17]

The satellite woes came at a time when American idealism was being shattered in popular culture as well. Over the preceding decade, television had transformed American home life. About 3.9 million U.S. households, less than one in ten, had a television in 1950.[18] By 1957, the figure had grown to 38.9 million of the nation's 49.5 million households, nearly four in five.[19] TV advertising expenditures reflected the growing power of the industry. Between 1950 and 1960, ad spending on television grew ninefold, from $170 million to $1.53 billion, propelling the medium past radio and magazines and fueling a trend toward newspaper consolidation.[20] But the TV business was rocked by scandal when it became public in 1957 that quiz shows—so popular they represented five of the top eight shows—were rigged.[21] Later in the year, the public also learned that record companies were giving kickbacks—"payola"—to radio disc jockeys.[22] With or without in-home television, every American could witness the Vanguard explosion via grainy, black-and-white Defense Department footage in a Universal-International newsreel titled "Satellite a Bust: Rocket Blows Up in First U.S. Try." Narrator Ed Herlihy, whose distinctive broadcast voice most people would recognize even if they don't know his name, laments, "What happened is already unhappy history—another setback for the United States in the race into outer space."[23]

But what popular media recorded as a disastrous, dispiriting launch failure was also a failure to manage public expectations, to appreciate the risks involved in live coverage of scientific experiments, and to keep political aspirations in line with technical progress. The launch was scheduled two months to the day after the Sputnik shocker. When the December 4 rocket firing was delayed multiple times—not uncommon in launch countdowns—and finally "scrubbed" for that day, many newspapers ran large, bold headlines across the entire front page announcing the postponement. American readers began to comprehend the complexities of modern rocketry in stories that described the launch vehicle's various components and blamed the delays on problems ranging from frayed wires to defective parts to gusty winds. "As Impatience Mounts, Fidgety Scientists Fuss with Bride-Like Missile," read a headline in the *St. Petersburg (FL) Times*.[24] While such stories helped educate the public about this new technology, the focus on problems and delays only reinforced the fact that the Soviets had already overcome these complexities. During the wait, the press corps scrambled for whatever story angles they could find—filing reports about how Cape Canaveral was chosen as the launch site, what the locals thought about the commotion surrounding the event, the odds of satellites colliding in space, and even interviews with each other. NBC cameraman Gene Barnes fretted to a print reporter that the longer the delay, the less likely he could get his film to a nearby affiliate and processed for airing on the evening news.[25] Camped on the beach miles from the launch pad, others complained there were too few updates from program administrators. "They keep referring to this as a test firing, but the public looks on this as THE satellite and deserves more information about it," said the *New York Times*'s Milton Bracker.[26]

This buildup to the launch contributed to an exaggerated letdown, made worse by the timing—the December 7 anniversary of Pearl Harbor. The *San Francisco News* even used the headline "Cold War Pearl Harbor."[27] *Time* magazine, in its December 23 issue, chastised the 127 members of the U.S. and foreign press corps who covered the launch, pointing out that few "gave any strong warning to editors and readers—as briefing officers warned them—that they were there for a test shoot, and that one of three missile tests turn out to be a flop-nik."[28] Scientists working on the Vanguard project had warned that the odds were against successfully placing a satellite into orbit on the first try. One told reporters it would be "a real miracle."[29] J. Paul Walsh, deputy director of Project Vanguard, said in a press conference before the launch, "We will be pleased people if it establishes an orbit, but we will not be despondent if it doesn't."[30] Those cautionary words did not cut through the swirl of anxiety and wishful thinking.

Lost in the coverage, in the hand wringing that followed, and in some historical accounts of the era is the fact that the original plans did not call for the rocket that burned to carry a satellite. Dubbed TV-3 for its status as a "test vehicle," it was a new rocket design, and the launch was the first attempt using three "live" stages, meaning all three contained fuel and would undergo a "burn" during flight.[31] Its predecessor, TV-2, used a live first stage but dummy second and third stages.[32] TV-2 was launched successfully October 23, but only after months of delays due to manufacturing problems and seven "static" (bolted down to prevent liftoff) test firings.[33] That the United States attempted to put its first satellite into orbit by placing it atop an unproven launch vehicle, with the eyes of the world watching, is a historical oddity that appears different in hindsight than it must have seemed at the time. Although the TV-3 explosion dominated public perceptions, the Vanguard program was an unqualified success; in a record thirty months it developed an entirely new space launch vehicle and successfully placed three satellites into orbit.[34]

Competing Programs

Project Vanguard began as a scientific initiative to launch a satellite as part of the International Geophysical Year (IGY). Proposed in 1950 and sponsored by the International Council of Scientific Unions, the IGY program was modeled after the International Polar Years, held in 1882–83 and 1932–33.[35] Astronomers were forecasting a period of increased solar activity from mid-1957 to the end of 1958, so the "year" of research actually spanned those eighteen months.[36] In addition to studying solar activity, scientists in sixty-seven countries participated in coordinated research in such fields as geodesy, geomagnetism, gravity, meteorology, oceanography, rocketry, and seismology.[37]

In the years leading up to its start, the United States and the Soviet Union proposed launching satellites during IGY, and by agreement, these "man-made moons" were to transmit radio signals on a predetermined frequency, allowing scientists around the world to track those achieving orbit. Calling orbiting satellites man-made moons seems quaint now, but in those days, headline writers needed a description that average newspaper readers could grasp, as shown in a United Press headline from February 15, 1956, in the *Sarasota (FL) Herald-Tribune*: "First Man-Made Moon May Be Visible in 1957."[38] The story introduces the term "artificial earth satellite" in the second paragraph and sticks with "satellite" for all but one reference through the rest of the story. Project Vanguard's director, Hagen, tells the reporter that to see the satellite, traveling from horizon to horizon in eight to twelve minutes at eighteen thousand miles per hour, "will take a little doing," even with binoculars.[39] According to the United Press story, the United States planned to launch ten satellites in all, the first being 30 inches in diameter and weighing twenty-one and a half pounds. That is larger than the 20-inch size that was ultimately decided upon, and much larger than the 6.44-inch "grapefruit" placed atop TV-3.

The shortcomings of trying to track a satellite visually, even using powerful telescopes, was a major factor in the selection of Project Vanguard as the IGY satellite program. After the U.S. National Committee for the International Geophysical Year decided, in early 1955, that the nation's participation should include launching a satellite, the Army, Navy, and Air Force put forth competing proposals for this chance to make history. All three had active rocketry programs, and interservice rivalry for priority and funding of weapons systems was intense. All three proposed modifying existing rockets, but only the Naval Research Laboratory proposed using one not tied to an existing military purpose—the Viking. The Army's Redstone and Air Force's Atlas rocket programs were part of the country's nascent intermediate-range ballistic missile (IRBM) and intercontinental ballistic missile (ICBM) fleets, and the Air Force had already begun developing plans for a military satellite to be launched using Atlas or Titan rockets.

The contrasting Army and Navy satellite proposals illustrate the trade-offs that early satellite designers faced in terms of size, weight, and functionality. A smaller, lighter satellite would be easier to put into orbit but hard to verify. If it achieved orbit, what further value could it offer to justify the effort and expense? A larger, heavier satellite could hold the instrumentation and batteries needed to transmit tracking signals and other data over a period of time, but lifting more weight high enough for a stable orbit stretched the capabilities of existing rockets. The Army proposed launching a small, five-pound satellite called Orbiter. Although the use of the well-established Redstone rocket promised an earlier launch date, the satellite as originally proposed lacked any means to transmit tracking signals and could perform no scientific experiments.[40] The Naval Research Laboratory's proposal, *A Scientific Satellite Program*, dated April 13, 1955, notes that such a satellite would be visible only at dawn or sunset, in favorable weather, and would be "exceedingly difficult to acquire in an optical instrument of sufficient power (and hence restricted field of view)" unless its location were already known.[41] "Indeed, it is readily conceivable that an object could be placed in an orbit and never observed, if only optical methods are used," the proposal warns.[42] To address this issue, eight pages of the proposal are devoted to describing in detail a tracking system called Minitrack, based on modifications to the guidance system used in the Viking rocket program. The modified Viking rocket and modified tracking system, Minitrack, together with a new, instrumented satellite design, became Project Vanguard.

Milton Rosen, chief engineer of the NRL's Viking rocket program, conceived Viking as a research tool to study the upper atmosphere, and at the time of the proposal, it held the altitude record for single-stage rockets—158 miles.[43] Rosen, an electrical engineer, had worked on guided missiles at the NRL as World War II ended.[44] First launched in 1946, Viking incorporated innovative "gimbaled" motors, which could be angled for steering the rocket, intermittent gas jets for stabilizing it after the main propellant was exhausted, and radio telemetry.[45] In his bid to win the IGY project, Rosen collaborated with Roger Easton, who had joined the NRL in 1943 and worked on radio beacons and blind aircraft landing systems, and Easton's boss, John Mengel, who headed the laboratory's Radio Division.[46] Mengel coined the name Minitrack from a phrase used in the title of a memo, "Proposal for Minimum Trackable Satellite," that he and Easton wrote to describe the system.[47] By switching to a lower radio frequency, which Rosen had suggested, and using large, five-by-fifty-foot ground antenna arrays, the system could pick up the relatively weak signal generated by a transmitter small enough to fit in the satellite. Minitrack used trigonometry to calculate the satellite's position by comparing the different angles of the incoming radio signal at pairs of ground antennas connected to receivers capable of detecting tiny differences in the signal wavelengths. As Mengel explained in a *Scientific American* article, humans use the same the technique to locate the direction of a sound that reaches their ears at different times.[48] For a visual illustration, think of sitting on a long, straight beach. Waves arriving perpen-

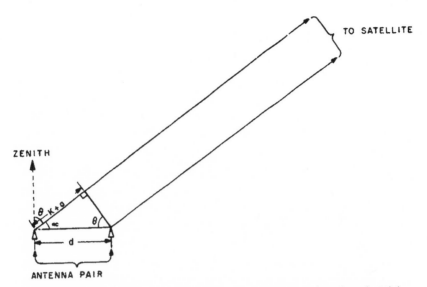

Fig. 1.2. Minitrack technique, using angles of signals. This diagram shows how the Minitrack system tracked satellites using the different angles of signals arriving at a pair of antennas. (Courtesy Naval Research Laboratory)

dicular to the shore break evenly, but those that roll in at an angle break from one side to the other and arrive at two points on the shoreline at different times. Figure 1.2 shows how the tracking technique was later illustrated in *Project Vanguard Report No. 18*, dated July 26, 1957, which was devoted solely to a progress update on Minitrack.[49]

The NRL's initial 1955 proposal goes into similar detail about the launch vehicle, describing modifications to the Viking rocket and the addition of two solid propellant stages. There is also a thorough discussion of orbital considerations, such as the advantages of launching eastward from a location near the equator to maximize the boost from the earth's rotation and the critical timing and precise angle needed in the final stage to achieve stable orbit. It provided fewer details regarding the satellite. The payload was to be "small in size and mass," between ten pounds and forty pounds—the projected weight of the shell, instruments, and batteries needed to transmit data for up to four months.[50] That is all that is stated—no materials, no dimensions, no schematics—but ultimately, once the satellite's design was fleshed out, it would join the tracking system as the winning elements of the proposal.

The NRL laid out an ambitious schedule of ten launches in thirty months, with three test flights and seven satellites, placed on the fourth through tenth rockets.[51] It predicted sending up the first satellite two years from the start of the program. Its main scientific objective was in the field of geodesy—using a satellite to make more accurate measurements of the earth, which would help in "tying together the various continental grids and locating the many islands with respect to these grids."[52] Geodesists relied on the moon as a reference point for making measurements in the middle of the ocean, and a satellite, by virtue of its closer proximity, would improve accuracy tenfold. In this case, a satellite truly would function as a "man-made moon." The data would yield practical and militarily significant results, the proposal notes: "Improved geodetic data is required to provide maps of sufficient accuracy for locating potential military targets and Loran navigation stations."[53] Loran, short for long-range navigation, was a land-based system of radio beacons operated by the Coast Guard. Left unstated is the reason the proposed partner in the effort, the Army Map Service, was interested in pinpointing several Pacific islands—they were to be used in ICBM test flights.

Beyond proving that a satellite had achieved orbit, other considerations factored in selecting the IGY satellite program. Among the concerns were how to keep the time and resources required from delaying existing military programs (no civilian organization had the wherewithal); how, for strategic and public relations purposes, to portray the launching of a satellite as a scientific rather than a military exercise; and how to keep sensitive military secrets from leaking out with published scientific findings. The National Security Council weighed these concerns, established policies for the program, and put the Department of Defense in charge of it in a secret directive, NSC 5520, adopted on May 26, 1955.[54] President Eisenhower approved the directive the next day.

Fateful Decision

Responsibility for deciding among the competing proposals fell to Assistant Secretary of Defense for Research and Development Donald Quarles, who asked an ad-hoc, eight-member selection committee to review the proposals. The panel included experts and scholars appointed by each branch of the military. Homer J. Stewart, an aeronautical engineer at the California Institute of Technology's Jet Propulsion Laboratory, chaired the committee, which came to be known as the Stewart Committee.[55] The group met in full or in part numerous times throughout the month of July 1955, even visiting the Martin Company factory in Baltimore to view rocket production.[56] President Eisenhower, however, did not wait for the Stewart Committee's decision before announcing that the United States planned to launch a satellite. On July 29 his press secretary, James Hagerty, met with reporters to confirm plans to launch "small earth-circling satellites" as part of IGY. His statement included this confident prediction: "This program will for the first time in history enable scientists throughout the world to make sustained observations in the regions beyond the earth's atmosphere."[57] The Soviets announced their IGY satellite plans four days later.

The Stewart Committee set aside the Air Force proposal, a pro forma effort that lacked a delivery date, seeing no way it could be accomplished without setting back progress on the Atlas ICBM. By July 1 the Army had revised its original proposal, modifying the upper-stage rockets and adding electronic tracking of the satellite. Faced with two alternatives, committee members split over which rocket held the most promise, while agreeing that the NRL's instrumentation and tracking were superior.[58] Committee member Clifford C. Furnas, a chemical engineer and guided missile expert who later succeeded Quarles as assistant secretary of defense, wrote about the deliberations in a *Life* magazine article, published October 21, 1957. Furnas said that the group wished it could have recommended using the Army's rocket with the Navy's instrumentation but realized from past experience with joint military programs that rivalry, jealousy, and unwillingness to share information, funding, and credit would so slow the process that getting a satellite into orbit by the end of IGY would be unlikely. "We finally decided that breaking the space barrier would be an easier task than breaking the interservice barrier," he wrote.[59]

In a five-to-two vote, with one member absent, the Stewart Committee selected Project Vanguard on August 3, 1955. The Army demanded, and got, a second hearing, as did the NRL. Vanguard's chief, Milton Rosen, offered revisions incorporating suggestions the committee had made—more tests of the new rocket and a lighter satellite—and provided written assurances from his four contractors that the schedule could be accelerated to launch the first satellite in eighteen months.[60] Rosen's original time frame would later prove far more accurate, as the program encountered numerous problems and launch dates slipped. However, the original committee decision stood, Quarles upheld the recommendation, and the Army was ordered to suspend satellite development. After Vanguard was selected, the Army's rocket men couldn't believe they had lost, and their counterparts at the Navy admitted surprise that they had won.

Considerable historical debate has focused on the Stewart Committee's decision. Some see anti-German bias toward the Army's team of former Nazi engineers, led by Wernher von Braun, which was responsible for the deadly German v-2 rocket.[61] Another viewpoint discerns partisan loyalties toward the Navy among the members of the Stewart Committee. Furnas, who voted with Stewart against Project Vanguard, later wrote that the decision was "purely technical. No politics were involved."[62] Still another claim is that the National Security Council's secret directive prohibited using a launch vehicle already intended for military purposes.[63] Motives are difficult to prove, but the text of the declassified NSC directive stops short of an explicit prohibition.

NSC 5520 is a short document, consisting of a dozen paragraphs and three attachments—a rough one-page budget, a discussion of technical matters, and an endorsement by Nelson Rockefeller, who was a special advisor to Eisenhower. The directive's thrust was to support spending about $20 million (Project Vanguard would end up costing $110 million) to launch a small scientific satellite, provided it did not impede research and development of long-range missiles or larger spy satellites the Pentagon was pursuing. When Vanguard's major contractor, Martin, won a contract to work on the Titan rocket, the best engineers who had worked on Viking were transferred to work on it. Thus, the reverse happened. Work on an ICBM impeded work on Vanguard. With the Soviet Union on the brink of developing intercontinental ballistic missiles capable of carrying warheads, reconnaissance satellites offered a potential means for keeping track of developments in that closed and secretive society. Unclear at the time, however, was the legal status of orbiting satellites. The directive authorized using a small, nonmilitary satellite to test the principle of "freedom of space." This doctrine held that national sovereignty of airspace did not extend to orbital altitudes, making a satellite akin to a ship in international waters. In an Oval Office meeting four days after Sputnik I passed over numerous nations without any protests, Donald Quarles, who had been promoted to deputy secretary of defense, observed that "the Russians have in fact done us a good turn, unintentionally, in establishing the concept of freedom of international space."[64]

Two months after signing the NSC directive, on July 21, 1955, Eisenhower proposed his "Open Skies" Treaty under which the United States and Soviet Union would have permitted reciprocal aerial reconnaissance flights.[65] He hoped to reduce the temptation to launch a surprise attack and to begin a process toward disarmament, but the Soviets rejected the proposal. Eisenhower authorized secret flights over Soviet territory, in violation of international law, and in 1960 the Soviets shot down Gary Francis Powers in a U-2 spy plane over their soil and a second reconnaissance airplane in international airspace. A few months later, the first CORONA reconnaissance satellite made further over-flights unnecessary.[66]

NSC 5520's language occasionally reflects the gap between what administration officials said in public and what was said in confidential meetings. The first sentence of paragraph 6 is revealing: "Considerable prestige and psychological benefits will accrue to the nation which first is successful in launching a satellite. The inference of such a demonstration of advanced technology *and its unmistakable relationship to intercontinental ballistic missile technology* might have important repercussions on the political determination of free world countries to resist Communist threats, especially if the USSR were to be the first to establish a satellite" (emphasis added).[67]

President Eisenhower publicly disavowed the idea that the United States was in a "space race" with the Soviet Union and in the aftermath of Sputnik projected an unworried response, playing golf that weekend (October 4, 1957, was a Friday) and waiting five days before holding a press conference. On the morning of October 9, Eisenhower released a prepared statement before fielding unusually contentious questions from the press corps.[68] His statement included these lines: "*Speed of progress in the satellite project cannot be taken as an index of our progress in ballistic missile work.* Our satellite program has never been conducted as a race with other nations. Rather, it has been carefully scheduled as part of the scientific work of the International Geophysical Year" (emphasis added).[69] Eisenhower defended the decision to keep the IGY satellite effort separate from military programs, asserting that if the United States had used a military rocket, it could have put a satellite into orbit before the Soviets, but doing so would have been "to the detriment of scientific goals and military progress."[70] Nevertheless, the day before the press conference Eisenhower authorized Army officials to begin preparing the Redstone as a backup for Vanguard.[71] They were more than ready.

Since the Stewart Committee's decision two years earlier, Maj. Gen. John B. Medaris, who commanded the U.S. Army Ballistic Missile Agency (ABMA) at Redstone Arsenal in Huntsville, Alabama, and his outspoken chief engineer, Wernher von Braun, had openly sniped at administration policies and covertly found ways to keep their rocket team in position to launch a satellite. When the Pentagon denied earlier requests for Redstone to be named officially as a backup program for Vanguard, they quietly found ways to keep their satellite aspirations alive within the Jupiter C rocket program, a successor to Redstone that played a critical role in developing warheads capable of surviving reentry.[72] Jupiter C soared 682 miles in altitude at a speed of 12,800 miles per hour for a distance of 3,355 miles on September 20, 1956, and could have placed a satellite into orbit if its fourth stage had been filled with fuel rather than sand.[73] Pentagon officials suspected von Braun might try to leapfrog Vanguard, and prior to launch they explicitly forbade Medaris from launching an unauthorized satellite into orbit. Some in Huntsville became convinced that Eisenhower wanted the Soviets to achieve orbit first.[74]

In late 1956 Vanguard director John P. Hagen and deputy director J. Paul Walsh were set to visit Huntsville to discuss placing the Vanguard satellite aboard an Army rocket.[75] Von Braun had made known he was not particular about which satellite his rocket carried. As Roger Easton remembered the episode, Hagen was about to leave for the supposedly confidential trip when a reporter called and asked if he was going to meet von Braun. A startled Hagen told the reporter no and canceled the discussions.

Medaris and von Braun maintained their readiness to launch a satellite as their warhead reentry program advanced.[76] Jupiter C achieved a milestone by launching the first object into space that was recovered intact, when a nose cone dropped by parachute into the Atlantic Ocean on August 8, 1957.

When the Army successfully launched America's first satellite, Explorer I, into orbit on January 31, 1958, on the first try, it was not the result of a miraculous 115-day crash program. Explorer I rode into orbit aboard a Jupiter C rocket renamed Juno I to reflect its nonmilitary purpose. And it was no coincidence that the satellite itself was not spherical but rather a bullet-shaped, eighty-inch-long, six-inch-diameter cylinder—a fourth-stage rocket casing. More than eighteen pounds of scientific instruments filled its upper half.[77] The gear included a spare radio transmitter built by the Naval Research Laboratory for use in a Vanguard satellite and scientific instruments provided by astrophysicist James Van Allen of the University of Iowa.[78] His equipment was originally slated for the Vanguard satellite, though it was also designed to fit in a satellite launched by von Braun's team. After Sputnik, it was officially transferred to the Army's program, and Explorer I detected the radiation belts surrounding Earth, which are named for Van Allen.

The lingering indignation over the Stewart Committee's decision is apparent in remarks Medaris made on the twentieth anniversary of the Explorer launch. Speaking to the Huntsville–Madison County Chamber of Commerce on January 31, 1978, the Reverend J. Bruce Medaris—he was ordained as an Episcopal priest in 1970, ten years after retiring from the military—called Vanguard "an ill-conceived, expensive, idealistic project which caused the United States to pay out a pretty penny to help that project reinvent the wheel."[79]

Cold War realities probably doomed Eisenhower's approach from the start. While he pursued a strategy of following dual military and civilian paths into space, reflecting the society he led, Khrushchev was under no similar constraints and saw the satellite effort as merely part of the arms race. His son, Sergei, later recalled that Khrushchev did not fully grasp the historic and scientific significance of the Sputnik feat until reading worldwide press coverage.[80] If Khrushchev initially failed to appreciate the propaganda value of being first into orbit, he lost no time capitalizing on it afterward. Soviet claims to have successfully tested an ICBM in August 1957 had drawn skeptical and dismissive responses, but Sputnik's radio signal, beeping on and off every three-tenths of a second at a much lower frequency than agreed upon for the IGY, had been heard by amateur radio operators worldwide. In the days following the launch, most Americans listened to the sound on radio and TV, and many watched the night sky trying to catch a glimpse of it. As an exclamation point, the Soviets announced what they called a new and "mighty" hydrogen weapon three days after Sputnik.[81] Fears surged that if Soviet missiles could deliver a satellite into orbit, they would soon be capable of carrying atomic bombs. When the news broke, Vanguard's technical chief Milton Rosen was in France with his wife, Sally. She later recalled that their French friends were worried about a Soviet attack.[82] Such fears were not unfounded, given implied Soviet threats against London and Paris the year before, when Britain and France supported Israel's invasion of Egypt following nationalization of the Suez Canal.[83] In the weeks following Sputnik, Nikita Khrushchev conducted a virtual victory lap and wound up as *Time* magazine's "Man of the Year" in the January 1958 issue, for having the biggest impact on world events.[84]

Raising Expectations

In the days following Sputnik, not only was Eisenhower grilled in public, he was getting an earful in private. C. D. Jackson, a former campaign aide and special assistant who served two stints as the president's speechwriter, offered a blunt assessment in a confidential memo dated October 8: "Within the past thirty days we have been treated to as skillfully executed an example of psychological warfare orchestration as I have ever seen."[85] Jackson was a psychological warfare expert, having helped craft Allied propaganda campaigns during World War II. He later led the organization that broadcast Radio Free Europe, served as a delegate to the UN, and held executive positions at *Time*, *Fortune*, and *Life* magazines.[86] He cautioned the administration to avoid the appearance of instituting a crash program to catch up. "If we are indeed geared to the IGY and have a schedule, the important thing is to stick to the schedule but make sure that when our satellite goes up, it goes all the way—and if not bigger than the Russians' let it be unmistakably better," he counseled.[87]

Unfortunately, in trying to implement this advice, Eisenhower publicly committed Project Vanguard to a decision made months before Sputnik, at a time when American scientists and politicians still assumed the United States would achieve orbit first and when the prospect of a launch failure did not carry the same risk of becoming a public relations fiasco. At Eisenhower's October 9 press conference, his prepared statement included the following: "In May of 1957, those charged with the United States satellite program determined that small satellite spheres would be launched as test vehicles during 1957 to check the rocketry, instrumentation, and ground stations and that the first fully-instrumented satellite vehicle would be launched in March of 1958. The first of these test vehicles is planned to be launched in December of this year."[88]

Five days after Sputnik, neither the press corps nor the American public seemed inclined to focus much attention on the distinction between a 6.44-inch "test" satellite and a "fully-instrumented" 20-inch satellite. Whichever satellite came first would be *the* satellite, *New York Times* reporter Milton Bracker observed to another journalist.[89] After the second Sputnik (carrying a dog named Laika, which died from heat exhaustion and was doomed in any case without a reentry vehicle), pressure to catch up with the Soviets increased, and the distinction undoubtedly blurred further.[90] Vanguard's Milton Rosen later told an interviewer, "If we were going to launch any rocket, it was going to have a satellite on it."[91]

Soon after Vanguard was selected for the IGY satellite effort, project officials added the two remaining Viking rockets to the Vanguard launch schedule, for a total of twelve launches. Recall that Rosen had promised the Stewart Committee additional test flights. Six, rather than three, were designated as tests and six were designated as space launch vehicles (SLVs)—meaning they would carry satellites. A chart in *Project Vanguard Report No. 1*, from January 1956, shows the first satellite launch, SLV-1, scheduled for sometime in October 1957.[92] To test tracking and guidance systems, the upper stage of the final test flight would include instrumentation but no satellite. *Project Vanguard Report No. 9*, published nine months later, specifies October 31, 1957, as the first satellite launch date, but over the following year, a variety of problems forced delays.[93] *Project Vanguard Report No. 20*, published in September 1957, reveals the launch date had slipped six months.[94] Offsetting the bad news is this optimistic note: "As a result of a successful test of the third-stage rocket aboard TV-1, it has become possible to revise the test program to provide complete Vanguard vehicle configurations earlier in the program. The test program heretofore called for a heavy instrumented nose cone on the third stage of TV-3; it is now planned to replace this with one of the 6.44-inch 6-antenna satellite packages and make TV-3 identical with TV-4."[95]

The first test launch, TV-0, was essentially a leftover Viking rocket whose primary purpose was to check out the launch facilities and tracking system, while TV-1 was a Viking rocket with a prototype of the new third-stage motor.[96] As stated earlier, TV-2 had a live first stage but dummy second and third stages, so TV-3 marked the first test of all new stages together. According to the NRL's Roger Easton, Air Force colonel Asa B. Gibbs, who worked with the Navy on Vanguard, gets credit for suggesting that small satellites be placed on the test vehicles. "If we are going to all this trouble, why don't we put a satellite up?" Gibbs asked.[97] He proposed the change and got it approved—months before Sputnik, rather than in response to it. Two additional test flights, using backup rockets for TV-3 and TV-4, were added to the launch plan. This created four potential opportunities to put a smaller test satellite into orbit before attempting the full-size version, but *Project Vanguard Report No. 20* assigns a "small probability" to achieving orbit and appears more focused on getting a good check of the Minitrack tracking system.[98]

It would be Sputnik, however, that provided the acid test for Minitrack. Martin Votaw recalled working long hours on paid overtime from the beginning of the Vanguard program—a pace dictated by the IGY schedule rather than competition with the Soviets. On the Wednesday before Sputnik's launch, the over-budget program circulated a memo barring additional overtime after that workweek ended. Votaw went home Friday looking forward to some rest and was having dinner when Easton called.

"They launched Sputnik," Easton said.

"Good, now we know it can be done," Votaw replied.

"You don't understand. We've got to track it."

"Can I eat supper first?"

"Well, yeah, but come back right afterwards."[99]

Votaw and other engineers worked around the clock for three days installing new antennas and modifying Minitrack to the lower frequencies the Soviets used. The overtime issue never came up again.[100]

Easton, Votaw, and others at the NRL spent months fabricating the 6.44-inch satellites, which had capabilities beyond the basic telemetry needed to test Minitrack. Transistors, still a novelty, were essential elements of what were then termed "subminiature" transmitters, but Easton had to send the first batch back to Bell Labs for replacement due to manufacturing defects.[101] An aluminum sphere was selected for its light weight, strength, reflectivity, and what they hoped would be uniform temperature. No one was sure what temperature a satellite in space would reach in direct sunlight. A glassy coating developed for the test satellites was later applied to the full-size ones as well. Easton came up with a method to detect the temperatures on the skin of the sphere and inside it by using separate transmitters centered on different frequencies, which varied slightly as the temperatures changed. One transmitter was powered by batteries, while the other was powered by solar cells attached to the outside. Easton carried a model of the sphere home to ponder how it could be secured under the nose cone and then released at the proper time. His result was a "chin strap," which encircled the orb and was attached by retractable pegs to a can-shaped device bolted to the top of the third stage.

Transistors and solar cells, highlighted in news stories generated during the launch delays, must have seemed quite advanced to the public—indistinguishable from a "fully-instrumented" satellite. And as we have seen, it was not the satellite but the rocket that failed on December 6, 1957. Unlike the Soviets, who kept their launch failures secret, the United States lacked the option suggested by the *Detroit Times*, which said after TV-3, "We should have kept mum until success had been attained."[102] With a free press and Project Vanguard funds flowing through the National Science Foundation, the administration foreclosed the option of total secrecy with its decision to launch the first satellite as part of the IGY program.

In retrospect, with these factors in mind, it would have been preferable in the wake of Sputnik to pull the small satellite from TV-3 and conduct additional rocket tests without satellite payloads to improve the odds of success under the glare of public scrutiny. Pressure to put a satellite in orbit seems to have blinded those in authority to the implications of any failure being so well publicized—precisely the concern C. D. Jackson expressed in his memo.

Vanguard's Legacy

Project Vanguard experienced another letdown when the TV-3BU backup rocket broke up fifty-seven seconds into its flight, on February 5, 1958.[103] Coming on the heels of the triumphant Explorer I launch five days earlier, the rocket failure brought a less intense public reaction. Success finally came on March 17, 1958, when Vanguard I, a "test"-size satellite, soared into orbit aboard the TV-4 launch vehicle. Altogether, over a period of about four years, Project Vanguard developed a new three-stage rocket, designed and built an accurate worldwide tracking system, and placed three satellites, weighing three, twenty-one, and fifty pounds, respectively, into orbit in fourteen launch attempts, eleven of which carried satellites. While the public's memory is forever tied to the TV-3 fireball, the program surpassed its original goals: to place a scientific satellite in an orbit around the earth, to prove that it was in orbit, and to use the satellite to conduct a scientific experiment in the upper atmosphere.

Although Vanguard I was not first into orbit, it exceeded expectations, starting with longevity. Sputnik I succumbed to orbital decay and burned up on reentry after three months.[104] Sputnik II lasted 162 days.[105] Explorer I transmitted signals for almost four months and orbited until 1970.[106] Vanguard I transmitted signals for more than six years and is still orbiting Earth in a highly stable orbit.[107] Predictions for how long it will remain in orbit range from two hundred to two thousand years. The satellite's long life, combined with Minitrack's accuracy, yielded surprisingly significant scientific findings, given its modest instrumentation. Orbital studies revealed, among other things, that Earth is not a perfect sphere but ever so slightly pear-shaped and that the upper atmosphere is far denser than previously thought.[108]

Longevity also describes Minitrack, which spawned several generations of ground tracking systems. By the fall of 1957, when Sputnik I reached orbit, Minitrack ground stations had been constructed at numerous sites around the globe, including at Blossom Point, Maryland; Fort Stewart, Georgia; Havana, Cuba; Quito, Ecuador; Lima, Peru; and Antofagasta and Santiago, Chile.[109] This north-south line of stations, located roughly along the seventy-fifth meridian of longitude and with overlapping, fan-shaped reception patterns extending skyward, created an "electronic fence" to capture each overhead pass of the satellite. Tracking stations transmitted all data by teletype to Washington DC for processing by an IBM 704 mainframe computer, which calculated and plotted each orbit.[110]

An astute reader may perceive here what scientists and engineers in those early days first comprehended—that if you could pinpoint a single radio-emitting satellite in orbit using multiple receivers on the ground, it should be possible by inverting the process to pinpoint a single receiver's location using multiple orbiting satellites—the essence of satellite navigation. However, the first operational satellite navigation system emerged from a different tracking method. A few days after Sputnik captured the world's attention, two physicists at Johns Hopkins University's Applied Physics Laboratory, George Weiffenbach and William Guier, calculated and eventually were able to predict Sputnik's orbits by analyzing the Doppler shift of its radio signal as the satellite circled the earth.[111] Doppler shift is the apparent change in pitch that occurs as a sound source moves by a listener. Trains, emergency vehicle sirens, race cars, and jets planes have made this effect so common that people today rarely give it a second thought. Frank McClure, chairman of the Applied Physics Laboratory (APL) Research Center, reviewed Guier and Weiffenbach's findings and challenged them (coincidently, the same day Vanguard I was launched) to "invert the solution," that is, to see if they could calculate a receiving station's position using a known satellite orbit.[112] They succeeded, and later in 1958 the APL's Richard Kershner led a federally funded project to build a system of radio-emitting satellites and worldwide tracking stations for Doppler measurements.[113] In 1964 the completed network became the Naval Navigation Satellite System, commonly called Transit. The military made it available for commercial use in 1967 (a pattern repeated later with GPS), and it operated until 1996, helping ships and submarines plot their position to within about five hundred feet anywhere in the world in any weather.[114]

The advent of satellites in orbit and dreams of sending humans into space led to the creation of NASA, the National Aeronautics and Space Administration, in 1958. Project Vanguard and most of the NRL people who worked on it were absorbed into the new civilian agency.[115] As satellites increased in size, number, and complexity, NASA reconfigured the network of Minitrack ground stations to keep pace. New polar and geosynchronous orbits (satellites traveling the same speed as the earth's rotation to remain at a fixed location above the equator) prompted the building of new sites and the closing of others, and Minitrack was renamed the Spacecraft Tracking and Data Acquisition Network.[116] It was followed by the Manned Spaceflight Network, also land-based, and both were replaced in the mid-1980s by the satellite-based Tracking and Data Relay Satellite System.[117]

Fig. 1.3. Frank T. McClure (*center*), director of the Research Center at the Johns Hopkins University Applied Physics Laboratory, chats with physicists William H. Guier (*left*) and George C. Weiffenbach. (Courtesy Johns Hopkins University Applied Physics Laboratory)

Roger Easton, not wanting to uproot his young family, decided to remain at the Naval Research Laboratory, where he turned his attention to one of the few space projects retained by the laboratory—tracking spy satellites.[118] The challenge of building tiny transmitters to fit inside satellites and accurate ground receivers to track them was replaced by the problem of tracking silent satellites designed to evade detection. As head of the Space Surveillance Branch, Easton led development of the Naval Space Surveillance System (sometimes shortened to NAVSPASUR, a moniker he never liked because it confused the system with its command structure of the same name).[119] The system still operates today as part of the larger Space Surveillance Network directed by the U.S. Strategic Command in Omaha, Nebraska.[120] Space Surveillance employs many of the same techniques used in Minitrack. A series of six ground receivers spaced across the southern United States from San Diego, California, to Fort Stewart, Georgia, forms a radar "fence," but the signals they track are generated not from satellites but by three large ground transmitters in Alabama, Texas, and Arizona. As satellites pass over the fence, continuous wave signals beamed into space by the transmitters bounce off the satellites and are picked up by the ground stations. A central computer at the program's headquarters in Dahlgren, Virginia, performs the high-powered computing required for quickly calculating orbits. When the system detected a large unknown satellite in late 1958, it created some excitement at the Pentagon, but the object turned out to be Vanguard's third-stage booster trailing behind the satellite.[121] The system's precision was sufficient to detect even the strap that had secured the Vanguard satellite to the rocket.[122] With improvements over the years, the system gained the capacity to detect basketball-size objects in orbit out to a range of more than seventeen thousand miles, and it is the oldest system for tracking the multitude of space debris now in orbit.[123]

The Space Surveillance system's continuous wave radar uses transmitters and receivers placed hundreds of miles apart. To measure the wave accurately, the receiver and transmitter must be precisely synchronized. Easton and his associates first tried transmitting the time code over the horizon, but extraneous noise introduced errors. Next, they carried an atomic clock by vehicle between the stations. That worked better but created a time-consuming task that had to be done continuously. In mid-1964 it occurred to Easton to put the atomic clock in a satellite, where it could transfer precise time to the transmitter and receiver simultaneously.[124] From there, using satellites with atomic clocks to transmit time precisely enough for navigational purposes, especially for the speeds associated with aircraft—something Transit could not do—seemed to Easton a logical next step. But the idea was not immediately embraced. "We were subjected to criticism because it was an idea looking for an application and not the other way around," Easton recalled in a November 2000 speech accepting a Distinguished Service Award at the Thirty-Second Annual Precise Time and Time Interval Meeting.[125]

If military leaders were not ready to launch atomic clocks into space, the clocks themselves were not quite ready for the mission either. Much work went into improving the accuracy of atomic clocks and "hardening" them to withstand cosmic radiation. Technological advances were needed in other fields, such as integrated circuits, which boosted the speed and shrank to a portable size the receivers used to process satellite signals. Complex computer simulations searched for the ideal altitude and arrangement of satellites to keep at least four in view at all times everywhere, while using the fewest possible to reduce cost. Engineers differed over the types of signals that would best transfer information from the satellites without being jammed. Funding was a perennial issue.

Gradually, advocates of advanced satellite navigation became more numerous, particularly in the Navy and the Air Force, and eventually at the Pentagon, which in 1968 decided the military could afford only one such system. A competition ensued, not unlike the one to launch the satellite for the International Geophysical Year, but this contest remained mostly unnoticed by the public, shrouded behind a cloak of military secrecy. That original competition persists today in differing narratives about the origin of GPS.

Subsequent chapters examine the competing claims, drawing on contemporaneous accounts published in technical journals, pertinent government documents that have since been declassified, and interviews with key individuals. Before looking at the competing designs that provided the building blocks of GPS, a short detour is in order. This was not the first time an advance in accurate timekeeping produced a revolution in navigation. For that story, it is necessary to go back a few centuries.

Weather Permitting
A Brief History of Navigation

2

Sleep did not fall upon his eyelids as he watched the
constellations—the Pleiades, the late-setting Bootes, and the
Great Bear, which men call the Wain, always turning in one
place, keeping watch over Orion—the only star that never takes a
bath in Ocean.

Homer, The Odyssey, *chapter 5*

The word *navigation* comes from *navis*, the Latin word for ship, and *agare*, the
Latin word that means to move forward.[1] Accurately knowing your position at
sea can make the difference between life and death. A prime historical example involves Sir Ernest Shackleton (1874–1922), who led several expeditions to
Antarctica. As World War I broke out in August 1914, he sailed from Britain on
the Imperial Trans-Antarctic Expedition. Its objective was to make the first
traverse of Antarctica. His ship, the *Endurance*, became trapped in pack ice in
the Weddell Sea, well south of the Antarctic Circle, and ultimately sank in
October 1915. After months on the ice floe, he traveled with his twenty-eight
men in three lifeboats to the inhospitable Elephant Island. From there, Shack-

leton made an epic voyage to South Georgia to summon help in the most sea-worthy lifeboat, the *James Caird*. His 850-mile route was through an area called the Furious Fifties, the zone between fifty and sixty degrees south latitude, where little land area interrupts the water circulating around the South Pole. Missing South Georgia meant certain death since no other inhabited islands were within reach. But taking accurate sightings of the sun is difficult on the rolling deck of a small craft in wind-tossed seas. When they approached South Georgia, his navigator, Frank Worsley, told Shackleton that he "could not be sure of our position within ten miles."[2] Given their exhaustion, the lifeboat's poor condition, and the danger of missing the island completely or being swept away, they headed for the nearest point of land on the uninhabited west coast of South Georgia. Landing successfully, they faced two difficult alternatives. The first was to sail around the island. Shackleton rejected that due to the poor condition of the James Caird. The second was climbing through the uncharted mountains and wastelands of South Georgia to the whaling station of Strom-ness. That had never been done, but Shackleton and the strongest two of his companions accomplished this feat, which even modern climbers with the best gear available find difficult.

Navigational Challenges

Prior to the balloon's development in the latter part of the eighteenth century, navigational challenges were mainly two-dimensional and could be separated into land versus sea travel. These two modes present similar but not identical navigational challenges. The navigator wants to safely travel from the current position, point A, to the destination, point B. Often, there are hazards to be avoided or interim points to be reached for resupply. Maps or sea charts pro-vide useful information, but two-dimensional maps distort the three-dimensional earth. And the earth is not a perfect sphere, which adds to the mapmaking and navigational challenge. An eighteenth-century French expe-dition to South America showed that Sir Isaac Newton was correct in his hypoth-esis that the earth is fatter at the equator than it is at the poles.[3] Gravitational variations exist in both the earth and the moon. One of the challenges in Apollo 11's landing was a lack of understanding of the moon's gravitational variations. This was intensively studied between Apollo 11 and 12, allowing Apollo 12 to make a pinpoint landing near the unmanned exploratory craft Surveyor 3.

Land travel has an advantage over sea travel in that there are more physical landmarks to help ascertain your position. Many landmarks are man-made, such as roads or cities. And repeated astronomical observations can be taken from a stationary land position, whereas an unanchored ship is always moving. However, navigating across a featureless desert is much like sea travel. Alexander the Great is believed to have visited the oasis of Siwa in Egypt. Danish scholar Torben B. Larsen gives a sense of how the ancients viewed the risks: "The overland journey was, according to the historian Callisthenes, a dangerous one. Alexander's party exhausted its water supply, but divine intervention produced a sudden downpour. A sandstorm caused them to lose their way, but divine intervention, Callisthenes says, sent two crows to lead them safely to Siwa."[4] Even today, mariners seeing nonmigratory birds know that land is nearby.

Every point on the earth can be specified by latitude and longitude. Latitudes are circles running east-west parallel to the equator, which is zero degrees. The North Pole is ninety degrees north; the South Pole is ninety degrees south. Longitudes run north-south, perpendicular to latitudes, but the starting point is arbitrary since there are no unique positions such as the equator or the poles. Given the British Empire's dominance in the nineteenth century, the British made the arbitrary decision to place the prime (zero) meridian through Greenwich in England. Using the previously defined navigational challenge as moving safely from point A to point B, the estimate of your current position on a map has some uncertainty, and the destination, point B, also has been measured with some uncertainty. During the Apollo 8 mission to lunar orbit, NASA found that the estimated position of the Command Module differed significantly depending on which ground stations were used. Managers at Johnson Space Center in Houston discovered errors in the coordinates of three remote island tracking stations. "The anomalous measurements from Canary, Hawaii and Guam were consistent with geodetic errors [in the positions of these stations] of up to 300 metres," recalled Pat Norris, a former Apollo navigation manager.[5] The distance that makes up a degree of latitude is constant whereas the distance for a degree of longitude is greatest at the equator and is zero at the poles.

North of the equator, latitude can be measured by the altitude of Polaris, also called the North Star, adjusting for the fact that it is not precisely over the pole. An alternative is measuring the sun's altitude at local noon, adjusting for the season. The sun lies directly over the equator at local noon on the first day of spring and the first day of fall. If the sun is forty degrees above the horizon at local noon on those days, a ship's navigator can calculate that he is at the fiftieth meridian. This method has limitations. North of the Arctic Circle, the sun cannot be seen at times during the fall and winter.

Longitude is much more difficult to determine. The British Parliament in 1714 passed the Longitude Act, which authorized a series of rewards for the person who perfected a method of determining it and established a commission, the Board of Longitude, to select the winners. The main competitors were clocks built by John Harrison, a self-taught clockmaker, versus various astronomical solutions.

Geometry has been used for millennia to measure distances. Eratosthenes, a Greek mathematician and founder of the discipline of geography, first calculated the circumference of the earth circa 240 BC. He knew that at noon on the summer solstice in Swene (modern Aswan, Egypt), a city located near the Tropic of Cancer, the sun was directly overhead, whereas in Alexandria the sun appeared at an angle south of the zenith. That angle equaled one-fiftieth of a circle. Eratosthenes reasoned, assuming that Alexandria was due north of Swene (it is actually slightly northwest), the earth's circumference is fifty times the distance between the two cities.

Harrison's timepiece, which developed into the marine chronometer, tells a mariner his longitude relative to the embarkation port based on its time compared to local noon (the sun's highest point in the sky). If the mariner's clock reads 1:00 p.m. at local noon, he is one-twenty-fourth of a day or fifteen degrees (one-twenty-fourth of a 360-degree circle) west of the embarkation port.

Accurate maps and charts are important for showing hazards. This is important even in the age of GPS. The ferryboat *Pride of Canterbury*, which travels between Dover, England, and Calais, France, hit the 1917 wreck of the ss *Mahratta* on January 31, 2008. The submerged wreckage was shown on the electronic sea chart but not at the magnification the *Pride of Canterbury* was using when the collision occurred.[6] Thus GPS by itself is not enough. One must also be aware of software idiosyncrasies.

Navigational challenges include finding a point along a coast or finding an island. Rear Adm. Daniel Gallery (1901–77) commanded a task force hunting German submarines in World War II. He seized the *U505*, the first ship captured by the U.S. Navy since the War of 1812, which now resides in Chicago's Museum of Science and Industry. In a novel written after the Six-Day War in 1967, he described current navigation techniques including Loran, a ground-based system of radio beacons, and the first space-based navigation system, Transit. However, his characters find Tel Aviv by sailing at a bearing where they can be confident that they will hit land north of the city. Then they could sail south and be confident that they would reach it.[7] Amelia Earhart, in 1937, failed to find Howland Island. Her navigator, Frank Noonan, wanted to fly to a point between Howland Island and Baker Island, which would have greatly increased their opportunity to check their position, but Earhart flew directly toward Howland Island and disappeared.

One myth is that ancient mariners hugged the coast. Amir Aczel, author of *The Riddle of the Compass*, points out that "the greatest danger a mariner faces is that of running aground."[8] Thus, it was critical to master navigation over open waters. In addition, more direct routes are generally quicker. Some methods Aczel discusses are sounding lines for measuring the depth of the sea bottom, knowledge of the shore profile, and knowledge of the winds, current, and the habits of various animals.[9] Geographical knowledge was codified over time with maps and sea charts. These gave sailors navigating across great distances interim points to check their accuracy and provided them with information about ports.

Navigational Techniques

Navigational techniques can be classified as celestial, mechanical, map related, and geographical. Celestial devices, as mentioned previously, use the sun, moon, planets, or stars. They are most commonly used to estimate latitude. The sun and the stars can also be used to locate east and west directions. Since astronomical objects that are not over the poles move from east to west, they also reveal the direction.

Mechanical devices were developed to improve the accuracy of these measurements. The astrolabe dates back to about 225 BC. Astrolabes were flat metal discs, typically about six to eight inches in diameter, made of brass, with markings representing astronomical features on a moveable faceplate. By rotating the faceplate and a watch-like hand to align with degree markings around the outer ring, users could determine the time of day or night and estimate sunrise, sunset, and the positions of stars.[10] Over time, more accurate devices were developed, such as the cross-staff, a simple T-shaped instrument with a moveable crossbar, used to measure the angle of the sun or stars above the horizon, and the octant, a triangular-shaped device with sighting mirrors at the apex and a curved bottom representing an eighth of a circle, or forty-five degrees, for determining latitude.[11] Royal Navy captain John Campbell invented the sextant in 1757. It improved upon the octant by increasing the curved scale to one-sixth of a circle, or 60 degrees, enabling measurements of larger angles up to 120 degrees.[12] A sky chart shows the relative latitude of the sun over the earth for each day of the year and is useful for determining latitude, if the sun can be observed at noon. The compass appears to have been invented in China around 800 AD. It allowed ships to sail year-round in the Mediterranean and greatly increased trade.

Lighthouses are useful for sea navigation. However, they are subject to dangerous misinterpretations, as occurred with the ss *Atlantic*. On March 31, 1873, the ship was headed with 814 passengers and 143 officers and crew toward Halifax, Nova Scotia, to replenish its coal. Lost in the dark, it hit a rock early the next morning. The officers and crew rushed to the deck and were able to free ten lifeboats, but the currents washed the boats away. Twenty people were killed on the deck when the bow on the foremast came loose. Ropes were brought ashore and some people managed to clamber across them to safety. Many married men refused to leave their wives behind and died with them. Five hundred forty-five people died in this catastrophe, which may have been caused by the crew confusing Sambro Light for Devil's Light, which was farther to the west. None of the 138 women on board the *Atlantic* survived; undoubtedly, they were handicapped by their heavy clothing. This included one woman disguised as a man who served as a crewmember.[13] It was the worst loss of life in a marine disaster prior to the *Titanic*.

Lunars, Jovian Moons, and Clocks

As mentioned earlier, the Board of Longitude was charged with finding a reliable way of estimating longitude. The largest award was £20,000 for a method that could determine longitude to an accuracy of half a degree of a great circle. The astronomical approach used celestial objects such as the Jovian moons or the lunar distances method, also called lunars. Galileo discovered four moons of Jupiter when he viewed the planet with a telescope in 1610. Their orbits range from 1.7 days for Io to 16.7 days for Callisto. Astronomers can predict their orbits and when they will pass behind Jupiter (be eclipsed by Jupiter). With four moons, there are many such eclipses. Observatories established tables showing the time of the eclipses at a well-known longitude such as London. Comparing the time on the published table with the time established by the observer's local noon gives the difference and the longitude compared to London. Measuring the local time of Jovian eclipses worked well on land; however, the eclipses could be seen only at night when the sky was clear and Jupiter was in sight. There are times in the year when Jupiter is on the opposite side of the sun from the earth.

Christopher Huygens, a leading seventeenth-century astronomer and mathematician, devised the first pendulum clock in 1656. John Harrison, as memorably recounted by Dava Sobel in *Longitude: The True Story of a Lone Genius Who Solved the Greatest Scientific Problem of His Time*, built on the work of Huygens and others in devising methods to offset the effects of heat and humidity on clocks. Harrison built increasingly sophisticated clocks in his attempt to win the £20,000 prize. His first clock, H1, was taken on a sea trial in 1736. It performed well, but the Board of Longitude insisted that a transatlantic trial was needed. His last clock, H4, a compact, five-inch-diameter sea watch, was taken by his son on such a trip in 1761.

The lunar method requires an accurate measurement of the moon's angle in relation to specific stars. Lunar tables were completed by 1767, the same decade in which Harrison's H4 received two official sea trials. Both methods were used by navigators between 1770 and 1850. Royal Navy captain James Cook took a replica of Harrison's H4 on his voyages of discovery and called it "our faithful guide through all the vicissitudes of climates."

The competition between the astronomical and the lunar approach ultimately came down to the price of the marine chronometer, since finding longitude with a chronometer was a faster and easier process and could be done every day provided you had clear weather. Frank Reed, who leads weekend celestial navigation demonstrations at Mystic Seaport, in Connecticut, spoke about lunars during the 2012 conference "After Longitude—Modern Navigation in Context" at the National Maritime Museum in Greenwich, England. He commented that the best time for "shooting" lunars is when the moon is half-full (first quarter or third quarter). It is done in daytime.

Inertial navigation, sometimes called dead reckoning, estimates position based on speed and direction. Thus a mariner on the equator sailing due west two hundred hours at five miles per hour ends up one thousand miles, or fifteen degrees, farther west since the circumference of the earth is about twenty-four thousand miles at the equator. During the first half of the nineteenth century, inertial navigation estimates could be adjusted for currents to make them accurate for about two weeks at a time. This worked well with lunar updates (adjusting for slight variations in the sun's path) every other week. Thus the longitude estimate from inertial navigation was updated by using the lunar method.

The chronometer, then costing about fifty pounds, was considered too expensive by American mariners. For example, the *Brig Reaper* voyaged in 1808 from America to Calcutta to buy coffee using inertial navigation and lunars. By 1850, chronometers became so inexpensive that the use of lunars disappeared.

The discovery of radio waves in the second half of the nineteenth century quickly became important to navigation. Scottish physicist James Clerk Maxwell mathematically predicted radio waves in 1864, and Heinrich Hertz, a physics professor at Karlsruhe Polytechnic in Germany, proved their existence by engineering instruments to transmit and receive them in 1887. Italian inventor Guglielmo Marconi and Serbian-American inventor Nikola Tesla both devised methods of wireless communication using radio waves. The British tracked the German fleet in the 1916 Battle of Jutland using radio direction-finding techniques. Radio engineers also recognized the feasibility of using radar for navigating ships or airplanes. Radar navigation is the reverse of tracking objects; the same reversal occurred in moving from tracking satellites to using them for navigation (see chapter 3).

The speed of airplanes accentuated navigational challenges, especially for solo pilots. There was no time for twenty-minute lunar calculations, and if there had been, the plane's speed prevented calculating a current position. U.S. Navy captain Philip Van Horne Weems pioneered methods for faster airborne navigation using mechanical devices. He invented a special watch for accurately determining Greenwich mean time, improved the sextant, and established a company, Weems & Plath, to market his navigational aids.[14] Along the way, he befriended prominent pilots such as Charles Lindbergh and the more controversial "Wrong Way" Corrigan, who claimed that he flew from New York to Ireland instead of Los Angeles due to misreading his compass. N. W. Emmot, in "The Grand Old Man of Navigation," an article based on an interview with Weems, notes that invoice records show that the information and charts Weems sold Corrigan were all about the North Atlantic. Another Navy aviator, Capt. Charles Blair, planned to fly in 1951 across the North Pole in a modified P-51 Mustang. Weems plotted in advance the sun's altitude for points along the entire flight, allowing Blair to compare his sightings to a graph without performing computations while flying.[15] In the 1960s, early position calculations using the space navigation approach utilized in GPS required extensive manual work. Today, miniature computer processors perform such calculations instantaneously on a smartphone or a dedicated GPS receiver. "The current almost universal utilization of GPS owes as much to advances in microprocessors as to progress in the central system itself," observed Dr. Alexander H. Flax, who served as chief scientist for the Air Force from 1959 to 1961 and as director of the National Reconnaissance Office from 1965 to 1969.[16]

Engineers at the Lorentz Company in Germany fielded the first electronic guidance system for low-visibility aircraft landings.[17] In 1940 the Germans sent bombers over Britain at night using two radio waves sent from the same station 180 degrees out of phase (peaks and valleys canceling each other out). When the bombers received no signal, it meant they were in the area where the waves overlapped and were headed toward the target city.[18] They also used a director beam, which the bomber would follow and then drop its bomb when it hit a cross beam.

Loran, for long-range navigation, was a system that developed out of the British Gee system, a land-based pulse radar that transmitted signals from towers in different positions. For a given time differential between the receipt of the signals, the receiver's position could be specified on a hyperbola. Thus, a receiver requires signals from another pair of stations in order to obtain a position fix. The system was improved when Loran-A was replaced by Loran-C, which operates at a lower frequency and is able to transmit signals over the horizon.

The next chapter covers the different space-based navigational system proposals in which satellites replaced land-based towers. After World War II, rocket pioneers in the United States and the Soviet Union built on the achievements of Wernher von Braun's German team, which developed the v-2 rocket. As noted in chapter 1, von Braun led the Army program that launched the first U.S. satellite. Space visionaries saw the potential for satellites to provide worldwide communications and navigation guidance unhindered by the "vicissitudes" of weather that have affected other techniques through the ages. Many people made navigation proposals; ultimately the Global Positioning System was formulated and revolutionized navigation. The potential problem today is overdependence on GPS, not finding uses for it.

Success Has Many Fathers
Early Concepts for Satellite Navigation

3

> Nothing in progression can rest on its original plan. We may as
> well think of rocking a grown man in the cradle of an infant.
>
> *Edmund Burke, letter to the Sheriffs of Bristol, 1777*

Controversies over priority in scientific ideas are common. There was debate
in the seventeenth and eighteenth centuries over the role of Sir Isaac Newton
versus Gottfried Wilhelm Leibnitz in inventing calculus. Newton invented it
first, but his aversion to publishing resulted in Leibnitz inventing it independently later and Leibnitz's notations proved to be more useful in practice. It
became a partisan issue, with the English asserting Newton's priority and Germans asserting Leibnitz's. There was a controversy in the nineteenth century
about the roles of British astronomer John Couch Adams and French astronomer Urbain Le Verrier in discovering Neptune. There was also the controversy
in the latter part of the nineteenth century over who invented the telephone,
especially between Elisha Gray and Alexander G. Bell. There were suggestions
in the twentieth century that Rosalind Franklin's role in discovering DNA's structure, for which James Watson and Francis Crick won a Nobel Prize, was overlooked because she was a woman. Note, though, that her early death made her

ineligible for the Nobel Prize, which is given only to living people. Similarly, a vigorous debate persists over the roles various people played in the origin of GPS.

Arthur C. Clarke, the author of *2001: A Space Odyssey* and a space visionary, wrote in 1945 about communications satellites in geosynchronous orbits. He anticipated a GPS-type system in a 1956 letter:

> My general conclusions are that perhaps in 30 years the orbital relay system may take over all the functions of existing surface networks and provide others quite impossible today. For example, the three stations in the 24-hour orbit could provide not only an interference and censorship-free global TV service for the same power as a single modern transmitter, but could also make possible a position-finding grid whereby anyone on earth could locate himself by means of a couple of dials on an instrument about the size of a watch. (A development of Decca and transistorisation.) It might even make possible world-wide person- to-person radio with automatic dialling. Thus no-one on the planet need ever get lost or become out of touch with the community, unless he wanted to be.[1]

Early Satellite Navigation Proposals

The following list shows eight American pre-GPS satellite navigation proposals.

U.S. Pre-GPS Space-Based Navigation Proposals

Angle
Edward Everett Hale, "The Brick Moon," 1870

Doppler
Lovell Lawrence Jr., "Navigation by Satellites," *Missiles and Rockets*, 1956
Transit, George Weiffenbach and William Guier, Applied Physics Laboratory, 1958

Range Measurement
Don Williams, Hughes Aircraft, 1959
SECOR (sequential correlation of range), Army, 1961
Roy Anderson, GE/NASA, 1963
Timation (time navigation), Roger Easton, Naval Research Laboratory, 1964
621B, Air Force/Aerospace Corporation, 1964

The list begins with one far predating the space age. Edward Everett Hale, more famous for his story "A Man without a Country," published "The Brick Moon" serially in the *Atlantic Monthly* in 1869. Hale proposed using satellites as an aid for measuring longitude, with one satellite over the Greenwich meridian and another one passing over New Orleans, both at altitudes of about four thousand miles.

The second and third systems used Doppler measurements. A year before the Soviets launched Sputnik, Lovell Lawrence Jr., an early rocket scientist who directed the U.S. Army's Redstone project, published "Navigation by Satellites" in the initial issue of *Missiles and Rockets*. He discussed placing satellites in geosynchronous orbits, but that was beyond existing technological capabilities. Geosynchronous orbits require altitudes high enough that satellites orbit the earth once each day, remaining roughly over one point on the earth. Such orbits require much effort to attain and maintain. A low orbit, six hundred miles in altitude, was more feasible at the time.

Sputnik's launch on October 4, 1957, led many people to track it. This was important for predicting Sputnik's orbit and studying how it decayed over time. Three individuals or groups who tracked Sputnik later proposed satellite-based navigational systems. This was no coincidence and warrants more attention than it often receives. Each recognized that the methods used for tracking could be transformed or inverted for navigation. Space tracking uses one or more ground stations to measure a satellite's orbit. Satellite navigation systems use one or more satellites to estimate a receiver's position.

As noted in chapter 1, three days after the Sputnik launch, William Guier and George Weiffenbach at the Applied Physics Laboratory at Johns Hopkins University listened to the satellite's twenty-megahertz signal and noted its pitch change over time due to the Doppler shift. They developed a space-based navigation system the following March using this shift. The first satellite launch, Transit 1A, in 1959, was unsuccessful. Transit 1B was successfully launched in 1960. The system used low-altitude satellites, at altitudes of about six hundred miles, and it entered service in 1964. It provided receivers two-dimensional position fixes periodically throughout the day. It was a useful system, especially for Polaris missile submarines, but over time it became clear that aircraft needed a three-dimensional, continuously available system, which would require a new approach.

Roy Anderson, a consulting engineer at General Electric's office in Schenectady, New York, followed Sputnik's track using radio direction-finder equipment set up in a camping tent. NORAD (North American Aerospace Defense Command) subsequently asked him to track each new satellite for forty-eight hours after launch.[2] He recalled the story of tracking Pioneer 4:

Three tracking stations demonstrated the ability to track Pioneer 4 to the great distance: Jodrell Bank in England with its 150-foot diameter antenna, the Jet Propulsion Laboratory (JPL) at Goldstone Lake, California with its 85-foot antenna, and a temporary setup at the GE Research Laboratory, Schenectady, with an 18-foot diameter parabolic antenna.... There was immense media interest in our effort. Pioneer 4 was the first object to escape Earth's gravity. We were besieged with phone calls at all hours of the day and night. With our small antenna we were seen as David against Goliath. On the morning of 6 March, the signal was weak and intermittent. Finally, search as we could, we could no longer get a lock on it. In mid-morning, a reporter called and said, "JPL announced that they lost a signal. Do you still have it?" "No." "When did you lose it?" "I don't know exactly?" "Can we say that at 10:25 you said that you lost the signal?" It was 10:25. "Yeah." By 10:27 the whole world was informed that GE had tracked the space probe farther than JPL. I went to a newspaper office and asked them to publish a disclaimer. They were not interested. JPL was not pleased.[3]

In 1959 Anderson discussed the use of range (distance) and range difference measurements for satellite navigation systems with Navy captain Alton W. Moody of NASA, who was president of the Institute of Navigation from 1959 to 1960. Range measurement computes the distance a radio wave travels from a satellite to a receiver, basing its calculations on the travel time, since radio waves move at the speed of light. Seventeenth-century astronomers noticed that observations of the Jovian moons were affected by whether the earth was on the same side of the sun as Jupiter or on the opposite side. This phenomenon showed that light travels at a finite speed. Nineteenth-century physicists hypothesized that it travels through a medium called ether. This reflected a philosophic aversion to the concept that space is largely a vacuum. Physicist Albert Michelson and chemist Edward Morely proved that the speed of light is a constant in a famous series of experiments in 1887. This disproof of the ether theory helped inspire Albert Einstein to formulate the special theory of relativity. Light travels at 186,282 miles per second. Consequently, the receiver is 18,628 miles from the satellite if the signal takes a tenth of a second to travel from it. Thus, one knows that the receiver is somewhere on the surface of a sphere 18,628 miles from the satellite.

Anderson submitted a satellite navigation proposal in January 1963 to NASA, which awarded him a contract. He proposed two modes of operation: an active

Fig. 3.1. Roy E. Anderson discusses satellite ranging with Richard L. Frey, James R. Lewis, and Axel F. Brisken (*left to right*) at the General Electric Earth Station Laboratory in Schenectady, New York, circa 1971. (Courtesy Gladys M. Anderson)

surveillance mode for transoceanic air traffic control and maritime safety and a passive navigation mode that would enable an unlimited number of users to determine their position from the satellite signals. There was more interest in the surveillance mode, since it was advantageous to know the position of commercial aircraft. For military applications, a passive mode is essential. In 1966 and 1967 NASA launched the ATS-1 (Applications Technology Satellite) and ATS-3 satellites, operating in the aircraft VHF band, which successfully tested Anderson's approach.[4] In 1964 Anderson proposed a navigation system with twenty-four satellites in four orbital planes with six-hour orbits at a fifty-one-degree inclination.[5] Ground stations would send time signals to the satellites, which then would relay them back down to receivers; thus, Anderson proposed a repeater system. Placing atomic clocks in the satellites did not appear feasible in the foreseeable future. Based on available evidence, Anderson was the first to propose a twenty-four satellite navigation system in a midaltitude configuration. These are important aspects of GPS today, and the selection of a passive design allowing unlimited users is a key reason for the system's success.[6]

Harold Rosen, who designed the first geostationary communications satellite at the Hughes Aircraft Company, lured to his team Harvard-educated Don Williams, who had proposed a space-based navigation system. Formulated in 1959, it called for two satellites in geostationary orbit for a range-measurement navigation system. Hughes vice president and later president Allen Puckett was participating in a yacht race from the West Coast to Hawaii, and Williams thought his navigation system could give Puckett an advantage. However, Puckett did not know about the proposal, and Williams did no further work on the navigation system after he began working on Rosen's communications satellite.[7]

Fig. 3.2. Roger L. Easton, former head of the Space Applications Branch at the Naval Research Laboratory, a few years before he retired in 1980. (Courtesy Naval Research Laboratory)

Roger Easton at the Naval Research Laboratory (NRL) joined the Rocket-Sonde Branch in 1952. This branch launched Viking sounding rockets—experimental research rockets that get their name from the nautical term for taking measurements—from White Sands Missile Range in New Mexico.[8] The NRL's Viking rockets should not be confused with the NASA project in the 1970s for a soft landing on Mars. Launching rockets in the 1950s was a chancy endeavor. Viking 8 in 1952, despite being bolted down, broke free and took off during what was supposed to be a static rocket test. Easton said that Milt Rosen, the head of the Viking project, looked as upset that day as any person he had ever seen. The 1955 Project Vanguard proposal that Easton cowrote with Rosen listed navigation as a beneficial use of satellites. It stated, "It would also be possible to determine the absolute longitudes and latitudes by observation of the satellite. Such observations would also yield the height of the observer above the center of the earth."[9] After the NRL's proposal prevailed over Werner von Braun and the Army's proposal as the official American satellite program in the International Geophysical Year, Easton designed Vanguard 1 and the Minitrack system to track satellites at the IGY-designated frequency of 108 megahertz. However, Sputnik transmitted at 20 and 40 megahertz, so tracking it required modifying Minitrack.

From Tracking to Navigation

After Sputnik's launch, Soviet spy satellites became a concern. These satellites would be silent, not emitting a signal most of the time; consequently Minitrack could not use their signal for space tracking. The new system Easton proposed in January 1958, Space Surveillance (formally the Naval Space Surveillance System, or NAVSPASUR), was an important development not only for its role in tracking satellites over coming decades. It created a technical problem— keeping the system's clocks synchronized—and Easton's solution of putting the synchronizing clock in a satellite led directly to his proposal for a navigation satellite system.

Space Surveillance worked as follows: "The system concept of NAVSPASUR is that of a continuous wave (CW) multistatic radar. A high-powered transmitter generates a large fan beam of energy, commonly called the 'fence,' which reflects signals from an orbiting object back to separate receiving stations. These receiving stations use large arrays of antennas as an interferometer to determine the angle and angle rates of arrival from the reflected signals. By observing the target satellite from several stations, the position can be determined; using multiple penetrations, the orbit can be inferred."[10] Thus the transmitter sends a beam into space; if it hits a satellite, the beam is reflected to receiver stations east and west of the transmitter. Easton's colleague, Martin Votaw, has called Space Surveillance "Minitrack with the transmitter on the ground rather than in the satellite."[11] Space Surveillance required receivers about one hundred times the size of Minitrack's.[12] Smaller satellites or those at higher orbits are more difficult to detect. ARPA, the Advanced Research Projects Agency, was established in 1958 in response to Sputnik to prevent tech-

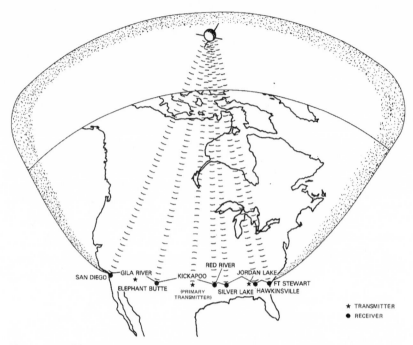

Fig. 3.3. Map of the Space Surveillance System. A series of powerful transmitters across the southern United States beam signals in the shape of a fan into space. Signals reflect back to receivers when a satellite crosses the beam. (Courtesy Naval Research Laboratory)

nological surprises. It approved funding for Space Surveillance in June 1958, and the initial satellite detection occurred in August 1958. This shows how rapidly the pace of technological change in space occurred after Sputnik. Navy captain David Holmes, who later played a major though unheralded role in GPS, pushed for installing a powerful, one-million-watt transmitter for Space Surveillance at Lake Kickapoo, Texas.[13] In 1960 work began on a second picket, or fence, in Texas, which turned the system into a radar and allowed it to estimate satellite orbits in one pass.[14] Space Surveillance transmitters and receivers were aligned in an east-west direction, whereas Minitrack receivers were aligned in a north-south direction. Early satellite launches aimed eastward to take advantage of the earth's rotation. A north-south system was optimal to detect these satellites as their orbits crossed overhead. For spy satellites, a polar orbit is preferable, because each orbital track passes over a new swath of ground as the earth rotates beneath it. An east-west tracking system is optimal for detecting these satellites.

Fig. 3.4. Capt. David C. Holmes at the U.S. Naval Academy in 1943. (Courtesy U.S. Naval Academy)

Peter Wilhelm, currently director of the Naval Center for Space Technology at NRL, worked in the Satellite Techniques Branch under Martin Votaw when he joined NRL in 1959. Wilhelm helped build the first electronic intelligence satellite, named GRAB. His first project for Roger Easton was building satellites to calibrate Space Surveillance. They were given the name SURCAL. Wilhelm asked if there should be an off switch for the signal, but Easton did not think it was necessary. The launch of SURCAL 1 was unsuccessful, but SURCAL 2 was launched in December 1962, and it jammed the system for about ten seconds on each orbit as its powerful transmitter overwhelmed the Space Surveillance system. Wilhelm said that it was a big irritant until it burned up three years later.[15]

The Timation (short for time navigation) program began in 1964. There is some uncertainty about the sequence of events that year. Synchronizing Space Surveillance's clocks in transmitter and receiver stations was critical for precisely obtaining the distance to a satellite being tracked. The tracking process relies on measuring the time it takes radio signals to travel from the transmitter to a satellite and be reflected back to receiving stations. If the clocks are not synchronized, an error is introduced. Since radio signals travel at the speed of light—299,792,458 meters per second—an error of one millionth of a second (1×10^{-6}) translates to an error of almost 300 meters. Easton recalled that in September 1964 he conceived using a clock in a satellite to synchronize the transmitter and receiver station clocks in Texas. In October he realized while on Naval Reserve duty that this would also make a good navigation system. However, other documents show that Timation's origin was more complicated. In a May 1967 slide presentation, Roger Easton stated, "To my knowledge the idea of using passive ranging as a navigation technique was a result of a conversation between myself and Dr. Arnold Shostak of ONR [Office of Naval Research] in 1964. Dr. Shostak was explaining how the hydrogen maser worked and obtained its fantastic time-keeping ability. At the time I remarked that this device appeared to make passive ranging feasible. He agreed and I spent a week working on the idea."[16]

He sketched more details in a June 9, 1964, memo showing how passive ranging works and discussing the accuracy of different types of atomic clocks. A July 1971 NRL chronology tracing the history of Timation states that it began in April 1964, so the conversation with Shostak probably took place then.[17]

It is not surprising that a discussion about atomic clocks between two Navy scientists led to a space-based navigation system using precise clocks. As shown in the preceding chapter, there is a historical linkage between maritime navigation and precise timekeeping. The chronometer was also known as the marine chronometer. Prior to the advent of aircraft, maritime travel was much faster than land travel apart from railroads, which followed fixed routes. The first atomic clock was an ammonia maser device built in 1949 at the U.S. National Bureau of Standards. Office of Naval Research funding supported the development of the first cesium atomic clock in 1951 and of the first hydrogen maser atomic clock in 1960. A paper Easton coauthored in the 1970s mentions that "in 1963, Study Group VII of the International Radio Consultative Committee referred to the advantages which might be expected to accrue from the use of time signal emissions from artificial earth satellites, and urged studies of the technical factors involved."[18]

Timation's initial test on October 16, 1964, used an NRL engineer's convertible. Matt Maloof drove along the unfinished Route 295 with a transmitter in his car, and people at NRL tracked his position.[19] Chester Kleczek, an engineer at Naval Air Systems Command, commented that Maloof was impressed that they could tell when he changed lanes. The results were sufficient for Kleczek to convince his boss, John Yob, to approve $35,000 for additional work. This was the largest amount Yob could give on his own authorization. The bureaucracy might have held up a larger amount, and it could have drawn opposition from proponents of Transit, which was also funded by Naval Air Systems Command.[20] The amount may seem small, even adjusted for inflation, but Easton commented that having an authorized program was important, and he could mix the seed money with other funds.[21]

Kleczek was born in Boston and studied at the Massachusetts Institute of Technology, but the Depression forced him to drop out after two years for financial reasons. Later he completed his degree at Northeastern University. As a sponsor for Timation at Naval Air Systems Command, he faced arguments in Pentagon meetings such as: Why do they need another navigation system when the government already supports forty-three navigation systems? Kleczek responded by asking which of them could give a Navy pilot his position over the South Pacific. There was a system, inertial navigation developed by Charles Draper, that could do it, but Kleczek knew that the people he was speaking with were unaware of it. However, inertial navigation could not be reset in flight if the power failed. Kleczek won approval to go to production.[22]

Kleczek feared that others would steal the idea for Timation and claim they invented it. Many people tried to find out how Timation worked, and Kleczek dodged their questions by telling them that NRL was still working on it. Kleczek recalled the period in a 2009 interview:

> I learned how APL worked. There was a satellite launch called Transit and they put out gravity gradient stabilization on a satellite and the damn satellite kept wiggling back and forth. They didn't know why. And NRL built one with gravity gradient but they put a damper between the satellite and the gravity gradient satellite, so there wasn't a strong coupling between the lower, the dummy, satellite and the real satellite and it's sort of slowed it down this way and it worked, it stayed vertical. Well, APL wrote a letter and they backdated it [saying] they had already invented it. I said, "Oh, you think we're going to tell them about Timation?" [Laughing.] That's what happened. But anyway that was just the skepticism that they would steal it and this was going on all over the place during those times.[23]

After the convertible test, NRL scientists simulated satellites using airplanes. An important breakthrough in winning support for continued research was explaining passive ranging using satellites in comparison to celestial navigation. Navy officers had for centuries navigated by measuring angles to stars. The Timation technique transformed this into measuring the time a signal took to travel from a satellite to a receiver. This explanation made sense to the Navy and facilitated its acceptance.[24]

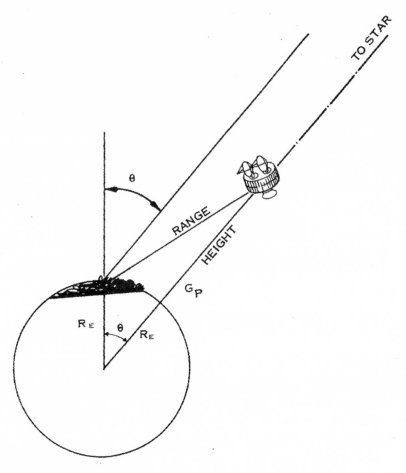

Fig. 3.5. Transforming celestial navigation to satellite ranging. This diagram helped to explain how navigation based on time signals from satellites compared to the angular sightings used in celestial navigation. (Courtesy Naval Research Laboratory)

NRL launched the Timation I satellite in May 1967 and conducted initial tests internally. Researchers gave Pentagon officials the first outside demonstration at the John Ericsson statue near the Lincoln Memorial in October of that year. They selected the location for several reasons. First, good sightlines were important, given that the satellite would be visible for only about thirteen minutes (it was in orbit about seven hundred miles above the surface of the earth). Second, Ericsson's invention of the screw propeller and the USS *Monitor*, the first ironclad battleship, made this an appropriate setting for demonstrating a major advance in navigation. Third, it was close to both NRL and the Pentagon. The successful demonstration yielded further funding. An article published to mark the fortieth anniversary of the test stated,

> The plan was to take one measurement at Time of Closest Approach [TCA] and one on either side between TCA and the horizon. As the data streamed in, [Don] Lynch called out which set of side tone readings should be resolved. Alick Frank then read the chart recorder deflections to James [Buisson, an NRL physicist], who resolved them and passed the result back to Lynch to be plotted on the intercept chart on the makeshift table. "The three points that Don chose were beautiful. They intersected very closely," Buisson reminisced later. After thirteen action-packed minutes, the satellite sank below the horizon and the pens fell silent.[25]

Easton wrote in the May 1967 slide show referenced earlier that constellation studies were deferred due to budget limitations but that both geosynchronous and midaltitude (mainly eight- and twelve-hour) circular polar orbits were being considered. Other tests were done with cars and boats, sometimes with unexpected occurrences. James Buisson and other NRL scientists were in a government van in Virginia, stopped on the side of the road waiting for the Timation I satellite to rise so they could track it. Buisson recalled, "A policeman stopped and asked us what we were doing. After we gave him a very detailed and long explanation about our atomic clock and the satellite, etc., he gave up trying to understand and said something like, 'Good luck on your experiment,' and he drove off."[26]

Different Needs, Different Systems

Rival navigation systems were proposed in the 1960s. The Joint Chiefs of Staff's Navigation Study Panel specified that navigation systems provide three-dimensional instantaneous position fixes worldwide within a specified accuracy. Accuracy in a satellite navigation system depends on many factors, including the accuracy of individual clocks, the synchronization between them, precise knowledge of each satellite's orbit, and corrections for changes in the signals caused by ionospheric distortion. Comparing two signals with different frequencies permits calculating and correcting for ionospheric distortion, so Timation II, launched in 1969, broadcast on two frequencies. NRL scientists extensively studied possible constellation configurations for Timation and determined that twenty-seven satellites, nine each in three evenly spaced orbital planes, with eight-hour orbits would be optimal. Ground stations were to be placed in Alaska, St. Croix in the Virgin Islands, Guam, and Samoa.[27] Security concerns limited ground locations to either the United States or U.S. territories. This restriction was eliminated in the 1990s after the end of the Cold War with the placement of stations in less secure areas, such as Diego Garcia in the central Indian Ocean. Eight-hour orbits were preferred to twelve-hour orbits since they provided more frequent clock updates and minimized the problems from not having a station in the Indian Ocean. Twelve-hour orbits became more feasible when atomic clocks replaced quartz crystal oscillators in the satellites, reducing the need for ground updates.

A widespread myth is that worldwide time synchronization was an unexpected consequence of GPS. "Certainly there was no serious consideration given to GPS becoming the de facto world standard for time," blogged Don Jewell in January 2008.[28] Jewell spent more than thirty years in the Air Force and writes for *GPS World*. A 2009 dissertation posited that, "somewhat unexpectedly, the precise timing information transmitted by GPS satellites was quickly incorporated into many inventive applications that are not related to navigation."[29] On the contrary, both Roy Anderson and Roger Easton anticipated worldwide clock synchronization. Anderson wrote in 1964 that his system would provide worldwide synchronization to approximately one microsecond.[30] Easton wrote in 1967 that "possible fallouts from such a system are worldwide time synchronized to better than 0.1 microsecond."[31] Thus the timing was more accurate by a factor of ten. In the *Timation Development Plan*, published in 1971, the estimated time transfer accuracy was improved again by a factor of ten, to better than 0.01 microsecond. Easton stated in 1974 that "precise orbiting clocks will prove to be a valuable tool in a variety of applications, by providing the entire planet earth with a single, accurate time system, enveloping the globe in a web of synchronized satellite signals."[32] This was about fifteen years before the term *World Wide Web* was coined. These satellite navigation proposals provided three-dimensional position and time. Even though time was less prominent than positioning, it was extremely important. Today, space-based navigation systems are often referred to as providing PNT, which means positioning, navigation, and timing. Time-transfer experiments via satellite began with the Timation I satellite in early 1968. Timation II performed the first international time transfer in July 1972 between the Royal Greenwich Observatory, England, and the Department of Defense Master Clock at the U.S. Naval Observatory in Washington DC.[33]

By the critical year of 1973, when the initial GPS system was configured, the proposed Timation configuration was twenty-seven satellites in three planes, in eight-hour circular orbits, with ground stations in the United States or secure U.S. territories.

The differences between the rival services' navigation proposals reflected their different needs. The Navy had a worldwide fleet, which included surface ships, submarines, and aircraft carriers. A worldwide three-dimensional system was important to meet the Navy's needs. The Air Force wanted to be able to put five missiles in the same hole. A worldwide system was not as critical a requirement for the Air Force.

The Air Force and its private research arm, the Aerospace Corporation, developed a space-based navigation system called Project 621B. Some sources assert that it started in 1963; however, a 1966 briefing report (declassified in 1979) by Aerospace engineers J. B. Woodford and H. Nakamura includes a chronology that dates "SSD/Aerospace identification of potential need for a new navigation satellite" to June 1, 1964.[34] The date is probably an approximation. The study envisioned regional constellations of geosynchronous or near-geosynchronous satellites and characterized putting atomic clocks in the satellites as a "growth item."[35] A paper presented by Woodford and two other engineers from Aerospace at the 1969 EASCON (Electronics and Aerospace Systems Convention) proposed three or four regional constellations around the globe, each with one satellite in geosynchronous orbit and three or four satellites in inclined elliptical orbits.[36]

The other 1960s range measurement system was the Army's SECOR, short for sequential correlation of range. SECOR satellites used transponders to return radio signals sent from three ground stations at known, surveyed positions to compute the position of a station at an unknown position.[37] It was designed primarily for geodetic purposes.

Toward the latter half of the 1960s there was pressure to consolidate space-based navigation system efforts and produce a system with three-dimensional capability that would be available worldwide and around the clock. Finding agreement on the design and deployment of a single shared system produced what Washington does best—multilayered committees, subtle infighting, budget wrangling, secret meetings, and afterward, differing accounts of what was accomplished and who should get credit.

One System, Two Narratives

4

Recollections and Documents

There are three stages of scientific discovery: first people deny
it is true; then they deny it is important; finally they credit the
wrong person.

Widely attributed to Alexander von Humboldt

Arguments about priority in inventing GPS are vigorous. Sorting them out is
challenging due to the loss or unavailability of important documents and the
effect of time on memory. Fortunately, they have not led to murder, unlike this
tragicomic story: "John Glendon of St. Clement-Danes, Gent. was tried for the
Murther of Rupert Kempthorne, Gent. on the 28th of October last, giving him
a mortal wound near the Navel, of the depth of 10 Inches, of which wound he
died the next day. The Evidence in general deposed, That the Prisoner and
Mr. Kempthorne were at the Ship-Tavern at Temple-Bar, and some difference
arose between them about Latitude and Longitude; Mr. Kempthorne alledg-
ing that there was no such word as Longitude."[1]

A Tale of Two Systems

The modern version of this argument concerns the importance of Timation, developed at the Naval Research Laboratory under the direction of Roger Easton, to the emergence and success of GPS. Some accounts of GPS history dismiss Timation entirely or sweep it aside as simply an effort to upgrade Transit, the Navy's first navigation satellite system.[2] That viewpoint traces the GPS lineage directly back to Project 621B, the navigation satellite program developed by the Aerospace Corporation, a federally funded research and development center sponsored by the Air Force.

Because the Pentagon appointed the Air Force as the executive service for GPS in 1973, and it has performed that role for decades, it is understandable that many assume GPS has always been entirely an Air Force program. Few people knew details about Timation during its development because it was a classified program, and its role became obscured once the NAVSTAR GPS program subsumed it. However, a reasonable examination of the individual characteristics of Timation and Project 621B, of the steps the government took to merge the two approaches, and of the resulting system, leads to the conclusion that GPS resembles the Timation approach more closely than that of Project 621B. The evidence that follows is not an attempt to discredit the contributions that Air Force managers, Aerospace Corporation scientists, or numerous private contractors made to the development of GPS. Rather, it attempts to present a balanced story, to promote wider appreciation for the contributions Timation made to GPS, and to accurately portray how GPS came to be the wildly successful program it is today.

Joint programs among the military services have always proved challenging and have sometimes failed. The F-111 fighter-bomber is an example of unsuccessful joint development of technology. Secretary of Defense Robert McNamara ordered the Navy and Air Force to develop the F-111 together. The Navy, judging the plane unsuited to carrier operations, was dissatisfied with its version, and the attempt to design an aircraft for both the Air Force and the Navy failed.[3]

As discussed in the previous chapter, various space-based navigation systems were proposed in the 1960s. The problems U.S. pilots had in the late 1960s and early 1970s destroying North Vietnamese bridges highlighted the need for precision munitions. But Vietnam War–era military budgets were stressed, so coordination between the military branches was encouraged. While the Pen-

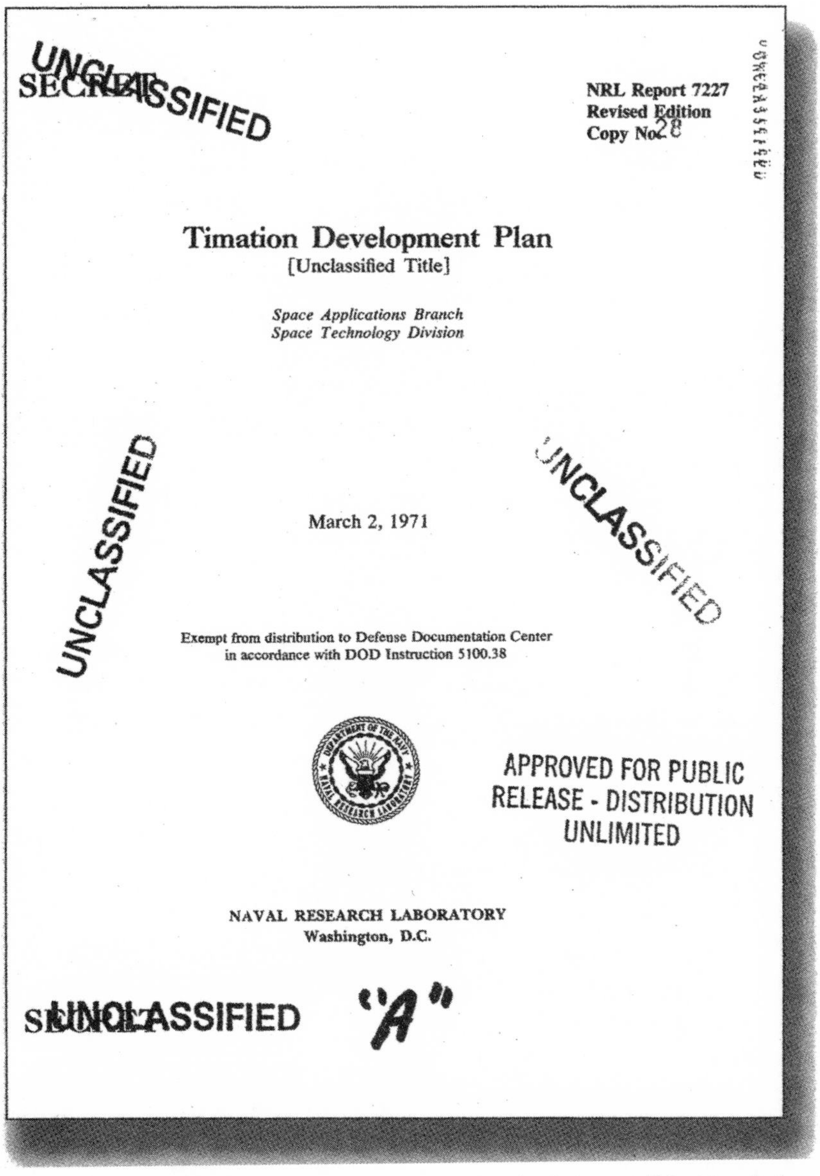

UNCLASSIFIED
SECRET

NRL Report 7227
Revised Edition
Copy No. 28

Timation Development Plan
[Unclassified Title]

Space Applications Branch
Space Technology Division

UNCLASSIFIED

UNCLASSIFIED

March 2, 1971

Exempt from distribution to Defense Documentation Center
in accordance with DOD Instruction 5100.38

APPROVED FOR PUBLIC
RELEASE - DISTRIBUTION
UNLIMITED

NAVAL RESEARCH LABORATORY
Washington, D.C.

SECRET UNCLASSIFIED "A"

Fig. 4.1. *Timation Development Plan*, 1971. The plan remained secret until the Navy declassified it in 1988. (Courtesy Naval Research Laboratory)

tagon had logical reasons to pursue a single, all-purpose satellite navigation system, differing needs among the services made joint development difficult.

The Department of Defense established the Navigation Satellite Executive Steering Group, or NAVSEG, in 1968. It was a tri-service group with the title of chairman rotating annually among the services. Harry Sonnemann was special assistant for electronics in the Office of the Assistant Secretary of the Navy (Research and Development) from 1968 to 1976. He was a member of NAVSEG from its founding to its dissolution and was its chairman from 1969 to 1970 and from 1972 to 1973. He states, "The Special Assistants responsible for oversight of Communications and Navigation Systems in the Offices of the Assistant Secretaries (R&D) of the Army, Navy including the Marine Corps, and Air Force, served as the Senior Members of . . . NAVSEG."[4] The Joint Chiefs of Staff established joint service requirements for a new space-based navigation system, including the ability for users to precisely position themselves in three dimensions and to precisely determine their velocity continuously, worldwide.[5] NAVSEG examined the services' different design schemes, which varied in the number of satellites, their altitudes above the earth, and their orbital inclinations (angle compared to the equator); the types of radio signals used; and the methods of controlling the satellites from the ground.

The 1969 Electronics and Aerospace Systems Conference mentioned earlier featured three papers advocating low-, medium-, and high-altitude satellite navigation systems. Three Aerospace Corporation engineers, J. B. Woodford, W. C. Melton, and R. L. Dutcher, delivered the paper "Satellite Systems for Navigation Using 24-Hour Orbits." Their preferred constellation—the one used for Project 621B—had one satellite in a synchronous equatorial orbit and three or four satellites in inclined elliptical orbits. Three or four such constellations could provide nearly global coverage. A master ground station and two or more calibration stations continuously tracked and sent time and position information to the satellites, which then retransmitted signals to the receiver. The system required the ground stations to be in the same area as the satellite constellations, since the synchronous satellite remains relatively stationary over a point on the earth. This was a major weakness for military applications. For example, the European constellation required ground stations in that area. A wag working on Timation in the 1970s commented that the 621B European constellation would have required its ground station to be in Moscow.[6] During a war, these stations would have been prime targets for direct attack or jamming the uplinks. A 621B satellite constellation could also have been destroyed by a single atomic bomb in space and the European constellation would have been denounced by the Soviets as a spy platform.[7] Even if the 621B satellites had atomic clocks, the system was more vulnerable than Timation since only ground stations in the same area could update the satellite clocks in synchronous orbits.

Roger Easton, in his EASCON paper "Mid-Altitude Navigation Satellites," stated, "After considering both lower and higher altitudes the mid-altitude (approximately one earth's diameter) polar circular satellite constellation has been selected as a prime possibility for an accurate, all weather, always available, three dimension, U.S. based navigation system." Later in the paper he stated, "The minimum number of satellites necessary to have three visible [from] anywhere on the earth's surface is approximately twelve." He then added that more satellites were desirable since while having three in sight is enough for navigation, a fourth is occasionally required to correct the receiver's clock.

Air Force Space Command senior historian Rick Sturdevant has written, "As early as 1969–1970, Aerospace Corporation president and GPS pioneer Ivan Getting had suggested to Lee DuBridge, President Richard Nixon's science advisor, that a presidential commission be created to review how satellite navigation ought to proceed, because there were so many potential users. After thinking about it for several weeks, DuBridge concluded that execution of Getting's proposal would be too difficult. He told Getting, 'there are too many people, too many bureaucracies, too much politics, and too many agencies involved. Why don't you just have the Air Force develop it the way we always did?'"[8]

Sonnemann comments, "It was our responsibility to arrive at a consensus with regard to the elements of the system, including the orbits of the satellites, the signal structure best suited for the project etc., etc., as well as management issues."[9] This was challenging due to the different objectives of the services. The Air Force focused on precision targeting of munitions, whereas the Navy needed worldwide navigation for its ships, including missile submarines near the North Pole, and for airplanes based on aircraft carriers.

NRL's Ron Beard, who was the second successor to Easton as branch head, has written that "from 1968 through 1970 the Timation concept grew from a Category 6.2 exploratory development project into navigation satellite techniques to a Category 6.3 development system concept."[10] One requirement for a Category 6.3 program was writing a development plan. Consequently, the assertion sometimes made that Timation was just a study plan is incorrect.[11] The March 1971 *Timation Development Plan* states, "The satellites provide the necessary data for the navigator to determine his latitude, longitude, altitude, and time. Not all navigators need to determine all these parameters. The user with the most stringent requirements will use four satellites in view and the equipment to receive signals from all four satellites; reduced requirements will result in one or more of the four signals being ignored by user's equipment. This system was configured within the JCS [Joint Chiefs of Staff] navigation accuracy requirements."[12]

Figure 4.2, from the same page of the plan, shows an airplane receiving signals from four satellites. This illustrates the most stringent requirement of three-dimensional position and time transfer. This is one of many primary source documents that refute the myth that Timation was a two-dimensional system that required an atomic clock in the receiver. Some people continue to assert the myth today.[13]

NRL's James Buisson, a physicist, and Thomas McCaskill, a mathematician, studied the optimal Timation configurations for the number of orbital planes, number of satellites, and orbital altitude. An important consideration was the placement of ground stations. For security purposes, the plan was to locate them only in the United Sates or secure U.S territories. Easton commented, "NRL found that the critical item in a satellite navigational system for unbelievable accuracies is the ground station location. The ground station location surprisingly determines the next item, the satellite constellation."[14] As mentioned earlier, a problem with Project 621B was its European constellation ground station. The U.S.-based restriction on Timation ground stations created a gap in the Indian Ocean where no station was available to update the satellites. A problem the people developing Timation faced was uncertainty about when atomic clocks would be developed that were sufficient to replace crystal oscillators (quartz clocks) in navigation satellites.[15] Timation I and II had crystal oscillators, and Timation III, scheduled to be launched in 1974, carried them in addition to two experimental rubidium atomic clocks made by the German company Efratom. Midaltitude satellite orbits, as is true for all orbits up to geosynchronous altitude, have the trade-off that higher altitudes give greater coverage whereas satellites at lower altitudes orbit more rapidly, permitting more frequent ground station updates. The June 1972 constellation study by Buisson and McCaskill proposed a three-by-nine constellation (three orbital planes with nine satellites each) at eight-hour orbits with evenly spaced planes.

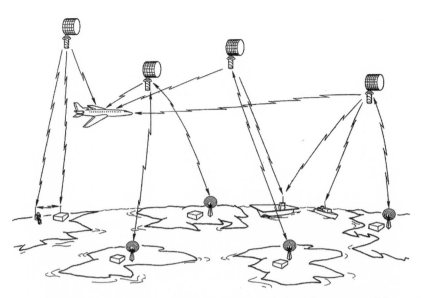

Fig. 4.2. Illustration of an airplane using Timation signals. This sketch from page 10 of the *Timation Development Plan* shows users on land, at sea, and in the air. The system was configured to provide position (three dimensions) and velocity accuracy to within fifty feet for strategic and tactical aircraft, when using four satellites. (Courtesy Naval Research Laboratory)

In April 1973 Deputy Secretary of Defense William P. Clements issued a memorandum creating the Defense Navigation Satellite Development Plan (DNSDP), a joint Army, Navy, Marine Corps, and Air Force program. He designated the Air Force as the executive service and directed it to assign a program manager and establish a joint program office (JPO). He listed the following guidelines for the program. Clements directed the Air Force to deploy during 1977 a constellation of four synchronous repeater navigation experimental satellites (NES). This was a 621B constellation. He directed the Navy to launch in 1974 the medium-altitude Navigation Technology Satellite #1 (renamed NTS-1; this was the satellite formerly designated Timation III). The secretary's memo requested a decision coordinating paper, a formal system design plan, by August 1973.

Navy captain David C. Holmes, a friend of John Glenn's, joined the Naval Research Laboratory in early 1973 as a consultant working on Space Surveillance and Timation. His important role in developing GPS has been largely ignored. A humorous story Holmes told gives one the sense of NRL's ethos:

They tell a great story about Dr. [A. Hoyt] Taylor [an electrical engineer who made important contributions to the development of radar] back in the early days of World War II. There was a rule at the laboratory that if you were the last person out of the laboratory you had to sweep up, and this applied to everybody. Dr. Taylor was sweeping up the laboratory one night in the early days of World War II when a young fellow came in. . . . He went in and saw this fellow sweeping out the place and asked him what kind of projects were done there. Then Dr. Taylor started telling him about the interferometers and various radar devices and things like that. He spent quite a bit of time explaining all these things. The young man went out talking to himself and saying, as he told the story later, "I certainly can't contribute to a place like that. Even the janitor knows more than I'll ever learn!" The young man's name was Arthur Godfrey [who later became a prominent radio and television broadcaster].[16]

Holmes, along with Easton, was a Navy representative to the tri-service meetings discussing the DNSS and had the contacts to convince the Navy to agree with the solution reached. A two-day meeting was held in June 1973 at the Space and Missile Systems Organization in California. Four alternatives were discussed for the DNSS (Defense Navigation Satellite System) and a draft decision coordinating paper (DCP 121) was written, dated June 7, 1973. Alternative I called for the Air Force to launch one NES satellite and the Navy to launch NTS-1. Alternative II was the same as the instructions in Deputy Secretary Clements's April 17 memorandum. During 1977 the Air Force would design and deploy a single constellation of four synchronous NES satellites and in 1974 the Navy would deploy NTS-1, a midaltitude satellite testing the Timation approach. Alternative III was similar to the second alternative, except that the Navy was to launch another satellite, NTS-2. NTS-1 would be launched in calendar year 1976 with a crystal oscillator to test signal processing in a subsynchronous eight-hour orbit. NTS-2 would be launched in 1977 with a cesium atomic clock. The fourth alternative was to proceed directly with a global, operational DNSS. No additional testing would be conducted prior to developing the satellites for launch in calendar year 1978. NTS-1 would be launched in 1974 to provide limited information on propagation effects (radio signal variability) and ephemeris (orbital variability), prediction methods required for the DNSS.

A Tale of Two Meetings

In August the newly founded joint program office, with Air Force colonel Bradford Parkinson as its head, attempted to get Project 621B adopted as the DNSS. However, the required approval from all of the services was not forthcoming. Parkinson claims that he and about twelve other Air Force and Aerospace Corporation personnel formulated GPS at the Pentagon in the "Lonely Halls" meeting over Labor Day 1973.[17] Roger Easton asserts that during the Labor Day weekend an important meeting occurred at a motel on Spring Hill Road in Virginia.[18] There, Easton and Captain Holmes, representing the Navy, met Parkinson and other Air Force representatives. Ron Beard and James Buisson from NRL recall Easton discussing the Spring Hill meeting the week after Labor Day. Beard recently questioned why Parkinson and other Air Force personnel from California would travel to the Washington DC area only to meet among themselves.[19] A meeting with people residing near the capital makes sense of their trip to the East Coast. Other evidence exists substantiating the Spring Hill meeting. Captain Holmes's daughter, Dian Moulin, found in his papers transparencies of a presentation about GPS's origins titled "Another Navigation System? Why GPS Wouldn't Sell." The last slide, reprinted in figure 4.4, corroborates Easton's recollections about the motel meeting.

Easton asserts that in the Spring Hill meeting, Parkinson said that 621B was rejected because it was too expensive and Holmes offered him the Timation system. Easton recalls that Parkinson listened but did not comment on the offer. Ron Beard recently wrote, "There were several meetings at the motel and elsewhere. I remember going to the motel also. The meeting that turned the tide was one that [Easton] and Holmes were at."[20]

Obviously, recollections of events that occurred decades earlier can honestly differ. Whether history should trace the compromise that led to the ultimate synthesis of GPS architecture to a meeting with two NRL representatives at a motel, to the Lonely Halls Pentagon meeting of Air Force and Aerospace personnel, or to its genesis over many months of tri-service negotiations depends on whom one asks. Sonnemann, who was part of those negotiations before the formation of the joint program office, offered this assessment: "In the short interval between the rejection of the 621B system by the DSARC in August 1973, the September 1973 Labor Day meeting, and the second DSARC in December 1973, there was not enough time to do any more than make sure that the elements of the system as then constituted by the NAVSEG/NAVSMO

Fig. 4.3. Col. Bradford W. Parkinson, first manager of the Joint Program Office, about a year before he retired from the Air Force. (USAF photograph)

[Navigation Satellite Management Office] were 1) Technically sound and 2) Would indeed satisfy the requirements that the synthesized Global Position System was the product of the NAVSEG joint service program and its supporting joint services NAVSMO staff."[21]

Keith McDonald, who headed NAVSMO, which provided technical and administrative support to NAVSEG, has written that the Air Force Space and Missile Systems Organization (SAMSO) "appropriate[d] the product of NAVSEG's concept development effort." In a telephone interview he observed that the final

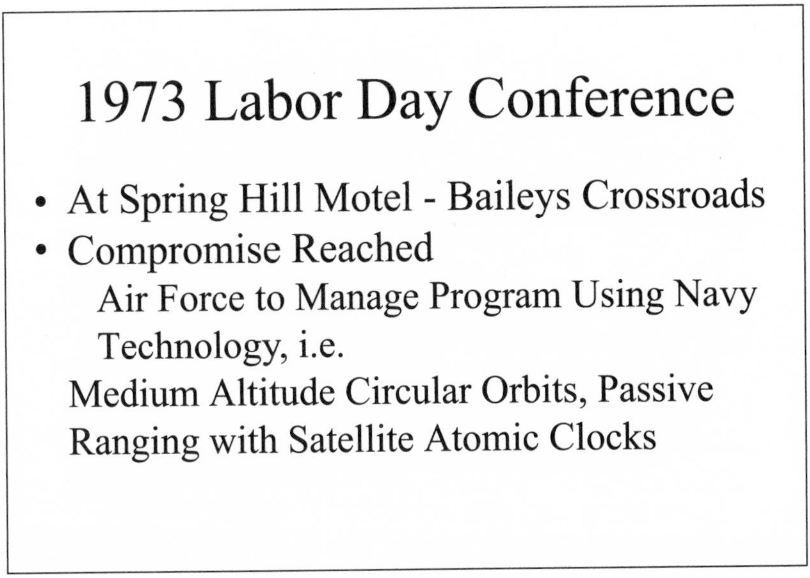

1973 Labor Day Conference

- At Spring Hill Motel - Baileys Crossroads
- Compromise Reached
  Air Force to Manage Program Using Navy
  Technology, i.e.
  Medium Altitude Circular Orbits, Passive
  Ranging with Satellite Atomic Clocks

Fig. 4.4. Holmes's presentation on GPS origins, slide 13. This reproduction of a transparency used in a presentation by Capt. David Holmes references a meeting at the Spring Hill Motel at Bailey's Crossroads in Virginia, where Navy representatives recommended that the Joint Program Office use the Timation system as the basis for GPS. (Courtesy Dian Moulin)

GPS architecture so closely resembled what NAVSEG members had already agreed upon, crafting the specifications would not have required extensive effort. "You could do it in half a day," he said. "It wouldn't take a weekend."[22]

Whichever viewpoint one accepts, it is clear that after the Labor Day 1973 weekend, the GPS system proposal reached a formulation with enough performance characteristics of the Timation system that Navy opposition ceased. The media reported on the compromise, and military officials touted it.

The main elements of compromise involved the signal, the onboard atomic clocks, and the orbital configuration. In a contentious exchange of letters to the editor of a technical journal in 1985, Parkinson responded, "As Roger Easton has noted, the Timation orbital configuration was, in fact, the basis for the Navstar system design."[23] But after the disagreement, Parkinson began asserting that the orbits GPS used were not the same orbits Timation used. Although the Block I experimental satellites were placed into orbits at a higher angle of inclination from the equator to facilitate testing and tracking, the Block II operational satellites used the fifty-five-degree inclination specified for Timation.

Parkinson often focuses on the techniques used to determine the precise locations of the satellites in their orbits, rather than the configuration itself, as he did in a 1999 interview:

> For example, Transit gave us orbit determination. They really knew how to do that, and we needed it, because GPS satellites have to know where they are. Roger Easton brought the atomic clock technology forward, and we needed very stable time because in essence GPS acts as a one-way radar. That is enabled by knowing precisely when the GPS signal was generated. The third contributions are the digital signal structure and the concept of operation. The Air Force was pretty close, and we took their signal structure and refined it further, added some features to it that weren't there. All these ideas contributed to the final GPS of 1973 (which is still essentially unchanged).[24]

This may overstate Transit's role in GPS. Two Timation satellites had already been launched, and NRL tracked their positions precisely enough to estimate the positions of receivers. Easton had designed two major space-tracking systems, Minitrack and Space Surveillance, so he knew how to track satellites. And the problems posed by gravitational variations are far less for satellites at higher, eight-thousand- or twelve-thousand-mile orbits than they were for the Transit satellites at roughly seven-hundred-mile orbits. "The GPS high altitude satellites will avoid the drag which continuously changes the orbits of the TRANSIT birds," Holmes wrote in an article published in the Naval Institute's *Proceedings* magazine.[25] *Aviation Week & Space Technology*, which closely covered satellite navigation progress, reported, "The ability of USAF and Navy to resolve their long-standing differences over the orbital configuration by basically adopting the Navy-proposed constellation arrangement has eliminated one of the major obstacles to Pentagon approval for the program."[26]

Three alternatives are presented in the subsequent Development Concept Paper Number 133, dated November 26, 1973. The first was to stop development leading to GPS. The second was to launch four synchronous repeater Navigation Development Satellites (a 621B constellation) and three NTS satellites. The third, which the paper's authors strongly advocated, was to launch two NTS satellites and three subsynchronous Navigation Development Satellites. This third option, with modifications, is what occurred. The 621B satellites were canceled, and the Timation satellites were launched under different names.

It is important to note that the *Timation Development Plan* specified using both sidetone (or continuous wave) signals and spread-spectrum signals (see chapter 5). This report was submitted on November 25, 1970. Thus the *Timation Development Plan* proposed using a spread-spectrum signal two years before Colonel Parkinson assumed leadership of Project 621B. This fact has been overlooked in debates about the origin of GPS. An August 6, 1971, memo from Air Force lieutenant colonel Paul S. Deem states that a System 621B signal modulator would be flown aboard Timation III (later launched as NTS-1). This occurred; thus there was cooperation between the Navy and Air Force long before the formation of the joint program office. There has been so much emphasis on the incorrect story that only the Air Force planned to use a spread-spectrum signal that even some of the people involved in Timation have forgotten the plan's intention to use both types of signals. Sonnemann, the former NAVSEG chairman, commented, "Once the orbit decision was made, a navy 'victory' in the eyes of the navy and the Air Force, the signal structure became the next hurdle. The use of the pseudo-random sequence structure proposed by the Aerospace Corporation for its 621B system had a number of attractive features, and no significant detrimental characteristics that would suggest the Timation signal structure should prevail. Adopting the 621B signal structure had the further advantage of balancing the slate by giving the Air Force a 'victory.'"[27]

Easton offered his assessment in a 1996 interview with Naval Research Laboratory historian Dr. David K. van Keuren:

Dr. van Keuren: Parkinson agreed after these meetings to take the Timation system and manage it [?]

Easton: Essentially.

Dr. van Keuren: Essentially. How much of the combined system was Air Force and how much was Navy?

Easton: I say, "What percentage do you want to put on it?" You could say it was ninety percent Navy and ten per cent Air Force, or you could use some other thing, but it was the Timation system with a different modulation. Essentially that is what it amounted to. That was in a way good because it gave the Air Force something to say that, well, it was a combination of two systems. The modulation, I thought, was the least essential thing because it could have been pulses FM, AM, sidetones, pseudo random noise, all kinds of ways you could have done it. We used the sidetones, and they worked, and we had hundreds of thousands of passes. . . . The original agreement was that we would have both, side tones and pseudo random noise. To me that made the most sense. You could have the sidetones for civilians and pseudo random noise for the military. If you needed to turn off the civilians, you just turn off that modulation.[28]

Former NAVSEG chairman Sonnemann was asked to provide his perspective on the elements that were synthesized to create GPS:

Question: In your estimation what percentage of the GPS system that emerged owes its technological origin to 621B?
Answer: The signal structure—100 percent.
Question: The same question for Timation?
Answer: The orbiting satellites (Timation) with precision atomic clocks—100 percent.[29]

However, the spread-spectrum signal structure was invented in World War II by the actress Hedy Lamarr and avant-garde composer George Antheil.[30] They deserve at least some of the credit for the signal used by GPS.

The Atomic Clocks

The Defense Systems Acquisition Review Council (DSARC) accepted the compromise on December 17, 1973. A major change was augmenting crystal oscillators with rubidium atomic clocks in NTS-1. Holmes described how that came about in an article for the defense industry magazine *Countermeasures*: "However, six months before the NTS-1 scheduled launch date a rubidium clock made by Efratom [a German company—the clock was developed by Ernst Jechart and Gerhard Huebner] was brought to the attention of Mr. Easton. Being an atomic clock, the rubidium oscillator offered the promise of both higher accuracy and higher, long-term stability. With joint program office approval, NRL decided to take the risk and install it, even though time available for testing and installation of space-qualified parts was minimal."[31]

For the second NTS satellite, Robert Kern and Arthur McCoubrey, who were leaders in developing atomic clock technology, built the first cesium atomic clock placed in orbit. NTS-2, launched in June 1977, provided useful information for GPS's further development, as reported at the time by *Aviation Week & Space Technology*: "Development test version of the Defense Department's NavStar global positioning system tracked an instrumented USAF/Lockheed C-141 with three-dimensional accuracies of 3–4 meters (10–13 feet) during a

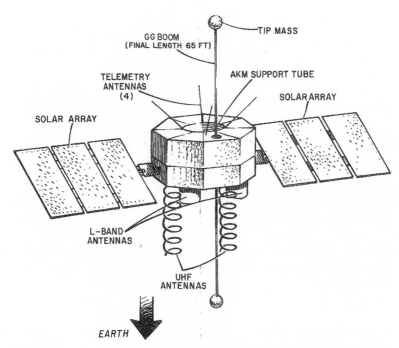

Fig. 4.5. Diagram of NTS-2 satellite. NTS-2, the final NRL navigation satellite, carried the first cesium atomic clock into orbit in 1977. (Courtesy Naval Research Laboratory)

recent test which synchronized the NavStar inverted range at Yuma Washington with the orbiting Navigation Technology Satellite (NTS-2). This constituted the first true navigational test of the NavStar-GPS concept, using a satellite and proved that the system is a workable one, according to officials from the joint program office at Air Force Space and Missile Systems Organization (SAMSO)."[32]

Harvard professor Peter Galison has written that Einstein's development of special relativity was at least partially inspired by observing synchronized clocks in Europe.[33] In turn, relativistic clock corrections are critical to GPS's provision of worldwide clock synchronization. NRL's Vince Folen and Don Lynch estimated correctly the relativistic corrections used in NTS-2's cesium clocks and University of Maryland professors Joe Weber and Carroll Alley were also very helpful in this calculation. The NRL's Peter Wilhelm and his staff built all four Timation satellites and developed innovative techniques to minimize launch costs. The 1971 *Timation Development Plan* proposed launching all twenty-seven satellites using three rockets with an estimated completion date of 1984.

After receiving approval in December 1973, Parkinson and the JPO imple-

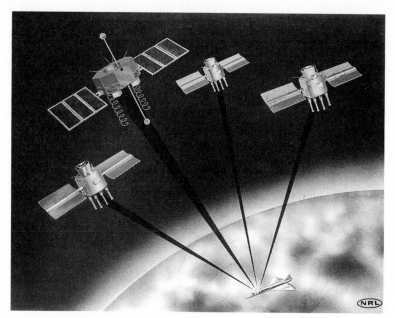

Fig. 4.6. NTS-2 satellite and first GPS demonstration constellation. An artist's rendering of NTS-2 and three subsequent navigation development satellites, which together created the first demonstration GPS constellation, giving users longitude, latitude, altitude, and time. (Courtesy Naval Research Laboratory)

mented the plan by working on the GPS Block I satellites and the receiver equipment. A major problem GPS managers continually faced was convincing some military leaders and many in Congress that this satellite navigation system's costs were worthwhile. Civilian applications would eventually be plentiful, but no civilian money supporting the program was forthcoming. Parkinson retired from the JPO and the military in 1978. Sonnemann commented about Parkinson's time leading the GPS program: "The Air Force JPO under Col. Parkinson did an excellent job in implementing GPS. Until the capability of the system was demonstrated and verified, no significant funds could be expected to be committed from non-DOD [Department of Defense] government organizations or the civilian potential users of GPS, so it would have been very difficult to speed up the process."[34]

In sum, when the documentary record and the recollections of those who differ with Parkinson's narrative are considered, a more complicated picture emerges. These sources reveal that the synthesis of competing satellite navigation approaches resulted from a long, complex interservice negotiation. That

Fig. 4.7. NRL NTS-2 team. The NRL team that launched NTS-2 included (*standing, left to right*) Dr. Bruce Faraday, Richard Statler, Guy Burke, and Roger Easton; (*seated, left to right*) Al Bartholomew, Cdr. Bill Huston, Red Woosley, Ron Beard, Woody Ewen, and Pete Wilhelm. (Courtesy Naval Research Laboratory)

process culminated in a compromise that incorporated the 621B spread-spectrum signal into a constellation of satellites carrying atomic clocks in a worldwide configuration of midaltitude orbits that resembles Timation, not the regional constellations of 621B. During the final, critical period of negotiations leading to the compromise, Parkinson appears to have played the role of an adroit broker, navigating the differing approaches of the two services into one aligned course. Although the two sides successfully merged the technical specifications into a single system, their stories about which system contributed each element and how they achieved the compromise remain divergent.

Invisible Stars
How GPS Works

5

Any sufficiently advanced technology is indistinguishable from magic.

Arthur C. Clarke, Profiles of the Future: An Inquiry into the Limits of the Possible, *1973*

Throughout history new technologies have intimidated those unschooled in their operation, and inventions sometimes have been branded as sorcery or witchcraft. This was true of early mechanical clocks; many eleventh-century "commoners" attributed their origin to the devil, so the introduction of clocks for general use flowed through the church, where monasteries found them useful for waking monks at prayer hour.[1]

Noted astronomer Herbert Hall Turner, former chief assistant at the Royal Greenwich Observatory and president of the Royal Astronomical Society from 1903 to 1905, once observed that to secure the Longitude Prize John Harrison "first had to disprove a charge of witchcraft by showing that his wonderful clock could be duplicated by another workman."[2] Since the Witchcraft Act of 1736 had repealed prior laws treating witchcraft as real and recast the crime as one akin to fraud or con artistry, the charge brought against Harrison was undoubtedly a tactic aimed at forcing him to divulge his secret design to the Board of Longitude.[3]

Despite the repeal of witchcraft laws, widespread ignorance, misunderstanding, and superstitions about science and technology persisted into the nineteenth century (some may argue that they extend to the present). In a magazine article first published in Charles Dickens's *Household Words* and reprinted in the inaugural 1850 edition of *Harper's New Monthly Magazine*, Frederick Knight Hunt noted "a superstition not wholly extinct" that the Royal Observatory in Greenwich was the "abode of sorcerers and astrologers."[4] In addition to telescopes the observatory featured a room full of chronometers being monitored while alternately subjected to extreme heat and cold—a test laboratory for the Board of Admiralty. Hunt suggested that the tendency to conflate astronomy with astrology was understandable, given that uneducated people might see little difference between predicting the future through horoscopes and producing almanacs foretelling "to a second when and where each planet may be seen in the heavens at any minute for the next three years."[5]

Capt. James Cook (1728–79) was not only a pathbreaking explorer, remembered for leading three famous expeditions around the world and visiting places no European had gone before. He was also an innovative seaman who introduced sauerkraut as a shipboard staple to prevent scurvy and experimented with distillation equipment to provide fresh water. Cook also was an expert navigator, surveyor, and mapmaker, having spent years mastering the technique of calculating positions using celestial observations and astronomical almanacs. Today we would call him an "early adopter" of new technologies. On his second voyage, from 1772 to 1775, he embraced a new technology that later came to be called the marine chronometer. Cook called it a watch, or watch machine. The timepiece carried aboard Cook's ship, the *Resolution*, was a replica of John Harrison's masterpiece, H4, the product of years spent designing a clock that would remain accurate at sea despite tossing waves and fluctuating temperatures and humidity. Clockmaker Larcum Kendall built the watch, so it acquired the nickname K1. A companion ship, the *Adventure*, carried three timepieces of a different design built by John Arnold.

The English Board of Longitude commissioned William Wales to join Cook's expedition as official minder of the Kendall watch, and Captain Cook's journals contain numerous references to Wales's observations comparing longitudinal positions predicted by the watch to celestial sightings and to established longitudes at inhabited islands. Early in the voyage Cook wrote, "Mr. Kendall's Watch thus far has been found to answer beyond all expectations."[6] About halfway through the expedition, in January 1774, Cook commented, "Indeed our error can never be great so long as we have so good a guide as Mr. Kendall's watch."[7]

A Giant Leap

While Captain Cook easily made the transition to using a mechanical time-piece for determining his longitude instead of measuring the distance between the moon and stars, the chronometer augmented celestial navigation; it did not replace it. Even Cook would have difficulty making the leap to our technology nearly two and a half centuries later. As a thought experiment, imagine transporting Cook through time to the bridge of a modern Coast Guard cutter for an introduction to its navigation equipment. His conversation with the commander might go something like this:

Cook: What a remarkable cabin you have here. What is its purpose? It cannot be where you sleep. I see no bed.

Commander: No, sir. This is the helm. We steer the ship from here.

Cook: You have built walls and a roof round your wheel? Where is the wheel? I don't see it.

Commander: We don't use a wheel anymore. We control the rudder with this. It's called a joystick.

Cook: I am at a loss to understand. How do the boatswains hear your line commands?

Commander: Well, we make announcements by speaking into this microphone and the sound of our voices comes out of speakers—like horns, you might say—located all over the ship. We also have bells, whistles, and lights, but we have no sails. An engine powers the ship, so the boatswains mates have other duties.

Cook: Engine? You mean the steam invention used for pumping water?

Commander: Yes, a descendant of that. They have advanced considerably. We don't use coal or steam. Diesel fuel—like lamp oil—powers our engine, which turns a ten-foot propeller.

Cook: Pro-pel-ler?

Commander: It's like a series of oars that turn in a circle—the same principle as Archimedes's screw. Instead of lifting water, this screw moves us through it.

Cook: (*shaking head and peering out the window*) You send a lieutenant on deck for your observations?

Commander: Coast Guard cadets still learn celestial navigation, but the fact is this equipment can provide all the navigation information we need.

Cook: What species of instrument is this? It looks like a small map on fine paper held up to a bright window.

Commander: That is our electronic chart display. It's a video screen—a picture that changes—connected to a computer—a calculating machine—that stores all of our maps and plots our course using GPS—the Global Positioning System. GPS can tell us our latitude, longitude, bearing and speed.

Cook: Indeed! The work of a sextant, a watch, a compass, and a log-line? You mean this machine can take observations of the moon and stars, calculate the distance, and search the lunar tables for you?

Commander: Not exactly. The information comes from the sky, but not from natural bodies. It comes from machines called satellites that are very, very high overhead—like tiny stars that orbit the earth, same as the moon.

Cook: Your language is quite new to me, and I am confused. It is midday. No stars are visible.

Commander: Well, these are too small to see, even at night.

Cook: What use are invisible stars?

Commander: We don't need to see them. The satellites transmit electromagnetic frequencies—um, radio signals that our equipment uses to determine our position. Radio signals are very rapid vibrations that our instruments detect with their antennas, which for them are like our ears, but these are not sounds anyone can hear.

Cook: You steer your ship with sounds you cannot hear from stars you cannot see?

Commander: I guess that is one way to describe it.

As this hypothetical conversation illustrates, understanding any form of technology relies on a base of knowledge about scientific advances that preceded it. The knowledge base—or at least an awareness of capabilities, if not a technical understanding of them—increases with each generation. Soon people take advances for granted. The principles of navigation using latitude and longitude have not changed, but Captain Cook would lack the knowledge base to readily grasp the workings of GPS. Meanwhile many casual users of GPS devices are ignorant about navigation methods, yet they readily adopt GPS technology. They take for granted electricity, radio waves, wireless communication, video screens, atomic clocks, spaceflight, satellites, solar power, microcomputers, and software programming. All of these are essential enabling technologies for GPS.

The Visible Parts

Of course, GPS satellites *are* visible, but because they orbit at an altitude of about 12,550 miles, seeing one requires a telescope. This distance is sometimes expressed as 20,200 kilometers, using the metric system favored by science and most nations, or 10,900 nautical miles, a traditional navigation unit based on the circumference of the earth. (A one-minute arc of latitude or one-minute arc of longitude at the equator is one nautical mile. This equals 1,852 meters, or about 6,076 feet, making a nautical mile 1.15078 times longer than a statute mile.) Regardless of the unit of measure, this distance amounts to roughly half the circumference of the earth, a comparison Captain Cook would have understood, or a bit more than one-and-a-half times the diameter of the planet, a more useful mental image for modern readers accustomed to viewing photos of the earth from space. Illustrations of the GPS constellation, such as the official image posted on the government's website (gps.gov; fig. 5.1), usually exaggerate the size of the satellites and trace invisible orbital paths but convey its overall scale.

The constellation contains several generations of GPS satellites with somewhat different designs. Each weighs around two tons, their heights range from about 6 feet to 11 feet, and their wingspans range from 17 feet to 116 feet.[8] The solar-panel wings charge backup batteries on board the satellites and power their clocks, signal transmitters, and other circuitry. While the height and weight of successive satellite designs have varied, the wingspan has grown steadily, raising the solar power generated from 800 watts to 2,450 watts.[9] Like the wingspan, the life span of the satellites keeps increasing. Engineers designed early satellites to last seven and a half years, and the newer ones have a twelve-year design life. However, many have lasted much longer than expected. Among older satellites designed to last seven and a half years, those remaining "healthy" at this writing include one launched in 1990, one launched in 1992, and six launched between 1993 and 1997.[10] The U.S. Coast Guard Navigation Center maintains a website (www.navcen.uscg.gov) where users can see the status of every GPS satellite. Each satellite has a unique space vehicle number, or SVN, but higher numbers are only generally indicative of age; the satellites

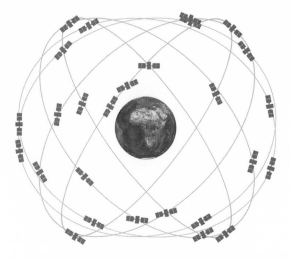

Fig. 5.1. GPS constellation, circa 2012. (Courtesy National Coordination Office for Position, Navigation and Timing)

were not launched in strict chronological order. Although the original system design called for a minimum of twenty-four satellites (including three spares) to accomplish all of its aims, surplus longevity has resulted in about thirty operational satellites since late 2004.[11]

The largest GPS satellite's silhouette is roughly the size of two semitrailers placed end to end. Anyone who has flown in a jetliner and viewed eighteen-wheelers on an interstate from five to seven miles overhead can appreciate how tiny they would appear from more than twelve thousand miles away. Many modern telescopes integrate GPS receivers into their aiming systems to speed input of the viewer's time and location, but spotting a GPS satellite, which must be in a position to reflect sunlight and is traveling at about 7,000 mph, remains a challenge.[12] Astronomers occasionally photograph GPS satellites and post their images online.[13] NASA's website features a tracking page where viewers can select from a list of satellites, including GPS satellites, to view animated, three-dimensional orbital paths.[14]

Each GPS satellite orbits the earth twice a day, or about every twelve hours. The earth is rotating at the same time, so the surface area covered by an individual satellite's signal constantly changes. To ensure that receivers can acquire signals from at least four satellites twenty-four hours a day anywhere on the planet, the satellites are divided among six orbital planes for better coverage. Each orbit is angled about fifty-five degrees from the equator, and the six orbits are spaced evenly around the globe, separated by sixty degrees of longitude where they cross the equator. Small rocket thrusters on each satellite allow ground technicians to maneuver new ones into designated "slots" along an orbital path, keep them aligned during use, and boost "decommissioned" satellites into higher orbits—the space-age equivalent of putting them out to pasture. Some of these "residuals" remain available for reactivation, if needed.[15] Technicians also keep each satellite's eight spiky antennas, called a helix array, properly aimed toward the earth to transmit signals.

As the number of satellites has increased and each generation's capabilities have improved, the system's ground control segment also has expanded and evolved. Since 1986, the Air Force's 2nd Space Operations Squadron (2SOPS) has managed the constellation from the master control station at Schriever (formerly Falcon) Air Force Base in Colorado.[16] An alternate master control station at Vandenberg Air Force Base in California is "functionally identical" and capable of assuming indefinite control of the constellation during any downtime at the master station.[17] A half-dozen original monitoring sites scattered around the world have grown to sixteen, and there are a dozen command and control antennas.[18] Integrating ten National Geospatial-Intelligence Agency ground stations into the control segment (a project begun in 2005 and completed in 2008) increased accuracy as much as 15 percent by placing each satellite under continuous active monitoring by three ground stations, whereas the satellites previously were unmonitored for portions of their orbits.[19] Continuous monitoring also enhanced security, offering faster detection of hostile attempts to disrupt the constellation.[20] Through this network of receivers and antennas, 2SOPS technicians monitor each satellite's health, compile and uplink almanac data containing details about each satellite's orbit (they wobble and vary slightly), and keep all operational clocks synchronized. Crews at the master control station generate an alert whenever any planned or unplanned outage or operational issue might affect signals. The Coast Guard Navigation Center website posts each alert as a Notice Advisory to Navstar Users (NANU), and users can sign up for automatic alerts. In 2007 the control segment's mainframe computer system, based on 1970s technology, was replaced with a modern information technology architecture. Upgrades to the operational control system continue.[21]

The Invisible (and Inaudible) Part

Unlike the Internet, with which users must connect to a server to download or upload information, GPS is a "passive" system in which users do not interact with the satellites or the ground stations. That means an unlimited number of users can simultaneously share the system by "listening" to broadcast signals, the same way any number of people can tune in to a radio station at the same time. It may be more accurate to say that a GPS receiver listens to multiple "stations," because depending on the receiver's design it may process signals from many satellites, both GPS and GNSS systems, such as Russia's GLONASS, Europe's Galileo, and China's Beidou, as well as signals from various land- and space-based augmentation systems. However, whereas a sound can be heard from around a corner, GPS signals are low-power "line-of-sight" communications. They can penetrate clouds, glass, and some thin fabrics or plastic but not most solid objects such as mountains, tree foliage, or buildings. This is why GPS literature so often features references to the number of satellites "in view."

Radio signals comprise content—music, speech, data, and so forth—and a carrier wave on which the content rides from the transmission source to the receiver. Radio waves are in the air all the time, but humans cannot hear them because their frequency is above the range of human hearing, like a dog whistle but much higher. Radio frequency is measured in cycles per second, also known as hertz (Hz), named for Heinrich Hertz (1857–94), a German physicist credited with discovering radio waves.[22] Kilohertz (kHz) and megahertz (MHz) describe frequencies in the thousands or millions of cycles per second, respectively. Humans generally hear sounds between 20 Hz and 20 kHz.[23] Dogs can hear sounds at about twice as high a pitch, around 40 kHz, and bats navigate by listening to echoes of sounds they project at frequencies around 120 kHz.[24]

The Federal Communications Commission (FCC) reserves a band of the radio spectrum between 88 MHz and 108 MHz for FM stations.[25] When a radio station operating at 100 MHz broadcasts a tone, say the musical note A above middle C on a piano, which vibrates at 440 Hz, the transmitter alters, or "modulates," the 100 MHz carrier frequency by 440 Hz. The listener's FM radio antenna receives the 100 MHz ±440 Hz signal, and an electronic filter extracts the 440 Hz tone to amplify through a speaker.

GPS signals ride much higher frequencies in a part of the radio spectrum called the L-band. Each satellite broadcasts signals on two channels, L1, at 1575.42 MHz, and L2, at 1227.6 MHz.[26] The content of these two broadcasts is digital codes—sequences of binary digits (zeros and ones) that feed information to GPS receivers. The signal structure utilizes techniques developed since the 1940s to spread transmitted information across a broad frequency bandwidth to gain signal-to-noise improvements.[27] Hence the name "spread spectrum." Spread-spectrum techniques are why FM radio (based on frequency modulation) is generally clearer than AM radio (based on amplitude modulation). GPS signals are quite weak after their twelve-thousand-mile journey, so spread-spectrum techniques help receivers pick them out of the cacophony of radio signals that fill the airwaves.[28] Military engineers utilize spread-spectrum techniques to make radio communications less susceptible to enemy interference (radio jamming is essentially noise) and to scramble signals so that enemy eavesdroppers perceive transmissions as noise.[29]

GPS uses a type of spread-spectrum code called pseudorandom noise (PRN) to digitize information about each satellite and the navigation data that receivers use to determine location. Each satellite broadcasts three PRN codes: the coarse acquisition (C/A) code, the precision (P) code, and the (Y) code, which replaces the P code whenever the military activates anti-spoofing measures designed to defeat intentionally misleading counterfeit signals.[30] When a user turns on a GPS device the receiver attempts to "acquire" or lock onto signals from as many satellites or other sources as its circuitry is designed to handle. Signals from a satellite directly overhead take about six-hundredths of a second to reach the ground. Receivers first detect the C/A code, which modulates L1 and repeats one thousand times every second.[31] This quick repetition helps receivers acquire satellite signals faster. The longer P code repeats on a seven-day cycle, making it harder to acquire. Because civilians use the C/A code and the P and Y codes are reserved for military use, many references speak of the two in combination as the P(Y) code. Most civilian receivers access only L1, while military users access both L1 and L2; accessing both helps to correct for atmospheric degradation. Otherwise, the accuracy of the civilian and military signals as they travel through space is the same.[32]

Each satellite transmits a unique C/A code. This identifier is different from the satellite's space vehicle number, so it appears separately, as the satellite's PRN code, on the Coast Guard Navigation Center's Constellation Status website. Another spread-spectrum technique, called code division multiple access (CDMA) arranges the digital sequences in a way that allows receivers to differentiate between numerous satellites transmitting similar codes on the same frequency.

Synchronicity

Aboard each satellite is an atomic clock, along with two or more backup clocks in case of malfunction. Extremely precise timing is essential for GPS accuracy, and a number of technological advances in timekeeping had to occur before satellite navigation could become a reality. Just as pendulum clocks were unsuitable for use at sea, the first timekeeping approaches that followed pendulums were not accurate enough to transmit precise time across the vast distance from a satellite in orbit to a user below. To appreciate how precise the time signal must be for GPS, consider that an error of one nanosecond—a billionth of a second—results in an error of about one foot.[33] An error of two microseconds—two millionths of a second—can send a speeding jet off course by a third of a mile.[34]

Pendulum clocks, with the addition of a second pendulum, reached their performance limit by the early 1920s, keeping accurate time to within ten seconds a year.[35] However, the growing radio industry, which uses frequencies in the millions of cycles per second, needed more precisely measured seconds to prevent one station's signal from drifting over another's.[36] The National Bureau of Standards (NBS; renamed the National Institute of Standards and Technology, or NIST, in 1988) searched for a material that would vibrate—oscillate—at a constant rate, like electrically driven tuning forks of the era, but much faster. The answer was to apply an electric current to a thin quartz crystal. Scientists learned to make quartz crystals vibrate at high frequencies that varied depending on the size and shape of the crystal. Quartz oscillators became the primary frequency standard at NBS, which set up four 100-kHz quartz oscillators that achieved accuracies of about one second in three years on average.[37] By 1950 quartz accuracy had improved by a factor of one hundred, and crystal oscillators replaced pendulum clocks at NBS as the primary standard for time intervals.[38] Quartz oscillators have some drawbacks. No two crystals are exactly alike or have identical resonant frequencies, and they wear out. Their frequency changes slowly due to aging and environmental variables such as temperature, humidity, and pressure.[39]

Scientists turned next to the atom. Physicists knew that atoms absorb or emit energy at specific frequencies, giving each atom a resonant frequency. Atoms represented an oscillator that would never waver or wear out, and all atoms of a given element would be identical. News of their potential use in clocks produced wonderment typical of atomic age discoveries. "'Cosmic Pendulum' for Clock Planned: Radio Frequencies at the Hearts of Atoms Would Be Used in Most Accurate of Timepieces," the *New York Times* announced on January 21, 1945, a day after Nobel Prize–winning physicist I. I. Rabi proposed the idea during a talk to the American Physical Society.[40] Seven months later the world discovered the awesome energy unleashed by splitting atoms when bombs destroyed Hiroshima and Nagasaki, but atomic clocks involve no fission or radiation. In 1949 NBS physicists announced they had synchronized a quartz-crystal oscillator to the natural resonant frequency of ammonia atoms. They did this by shooting high-frequency microwaves of about 24 billion Hz (24 gigahertz [GHz]) through a thirty-foot coiled copper tube filled with ammonia gas, causing each ammonia molecule's single nitrogen atom to flip its position among three hydrogen atoms, like an inverted pyramid.[41] They could not see this happen; they measured the effects. At frequencies too high or low, the microwaves passed through the chamber, hitting a detector that adjusted the frequency until at 23.8 GHz the gas absorbed the microwaves. This ammonia device was accurate to within one second in eight months.[42]

Atomic clocks (scientists prefer the term *atomic frequency standard*, or AFS) got smaller and more accurate over time. Cesium, a mercury-like metal that liquefies just above room temperature, replaced ammonia as the preferred atomic frequency standard.[43] Instead of shooting microwaves through a cloud of molecules, second-generation atomic clocks shot a beam of vaporized cesium atoms through a microwave signal and compared the energy they contained to another stream that was diverted around the microwaves by magnets. In 1952 a cesium clock achieved an accuracy of one second in three hundred years—for the first time dividing a second into billionths, creating the nanosecond.[44] Cesium clocks later achieved an accuracy rate of one second in twenty-five hundred years, and in 1967 the standard international second was redefined as the resonant frequency of cesium atoms, replacing the astronomical definition based on a fraction of the solar day.[45] In the 1960s Hewlett-Packard developed a line of 180-pound cesium clocks portable enough to transport in planes to synchronize timekeeping facilities around the world. It was about this time that Roger Easton at the Naval Research Laboratory envisioned placing those clocks in the air continuously aboard satellites.

The GPS constellation at this writing uses five cesium and twenty-six rubidium atomic clocks.[46] Rubidium is a silvery-white metal that is easy to vaporize. Clocks using rubidium in newer GPS satellites have demonstrated superior performance and longer lives than cesium clocks.[47] Although newer clocks are superior, the military conservatively rates all clocks in the constellation to be accurate to within one hundred nanoseconds over any three-second interval.[48] Accuracy on the ground depends on a variety of factors, including atmospheric

$$\begin{bmatrix} N & \sum_{i=1}^{N}\frac{\partial(R_c)_i}{\partial x_s} & \sum_{i=1}^{N}\frac{\partial(R_c)_i}{\partial y_s} & \sum_{i=1}^{N}\frac{\partial(R_c)_i}{\partial z_s} & 0 \\[2mm] \sum_{i=1}^{N}\frac{\partial(R_c)_i}{\partial x_s} & \sum_{i=1}^{N}\left(\frac{\partial(R_c)_i}{\partial x_s}\right)^2 & \sum_{i=1}^{N}\left(\frac{\partial(R_c)_i}{\partial x_s}\right)\left(\frac{\partial(R_c)_i}{\partial y_s}\right) & \sum_{i=1}^{N}\left(\frac{\partial(R_c)_i}{\partial x_s}\right)\left(\frac{\partial(R_c)_i}{\partial z_s}\right) & 2x_s \\[2mm] \sum_{i=1}^{N}\frac{\partial(R_c)_i}{\partial y_s} & \sum_{i=1}^{N}\left(\frac{\partial(R_c)_i}{\partial x_s}\right)\left(\frac{\partial(R_c)_i}{\partial y_s}\right) & \sum_{i=1}^{N}\left(\frac{\partial(R_c)_i}{\partial y_s}\right)^2 & \sum_{i=1}^{N}\left(\frac{\partial(R_c)_i}{\partial y_s}\right)\left(\frac{\partial(R_c)_i}{\partial z_s}\right) & 2y_s \\[2mm] \sum_{i=1}^{N}\frac{\partial(R_c)_i}{\partial z_s} & \sum_{i=1}^{N}\left(\frac{\partial(R_c)_i}{\partial x_s}\right)\left(\frac{\partial(R_c)_i}{\partial z_s}\right) & \sum_{i=1}^{N}\left(\frac{\partial(R_c)_i}{\partial y_s}\right)\left(\frac{\partial(R_c)_i}{\partial z_s}\right) & \sum_{i=1}^{N}\left(\frac{\partial(R_c)_i}{\partial z_s}\right)^2 & 2z_s \\[2mm] 0 & 2x_s & 2y_s & 2z_s & 0 \end{bmatrix} \begin{bmatrix} \Delta K \\ \Delta x_s \\ \Delta y_s \\ \Delta z_s \\ \Delta g \end{bmatrix} = \begin{bmatrix} \sum_{i=1}^{N}(1)(O-C)_i \\[2mm] \sum_{i=1}^{N}\frac{\partial(R_c)_i}{\partial x_s}(O-C)_i \\[2mm] \sum_{i=1}^{N}\frac{\partial(R_c)_i}{\partial y_s}(O-C)_i \\[2mm] \sum_{i=1}^{N}\frac{\partial(R_c)_i}{\partial z_s}(O-C)_i \\[2mm] 0 \end{bmatrix}$$

Fig. 5.2. Instantaneous navigation math. Mathematicians at the Naval Research Laboratory formulated this solution to the problem of two-, three-, and four-dimensional instantaneous navigation using passive ranging signals from satellites. (Courtesy Naval Research Laboratory)

effects and receiver quality, but the accuracy of the Standard Positioning Service offered to civilians is generally three meters or less—under ten feet—99.99 percent of the time. Various techniques used today, including augmentation systems discussed in later chapters and high-end receivers that measure the carrier wave itself, can improve that accuracy to a few millimeters.[49]

Before GPS receivers display latitude, longitude, and altitude figures or a moving dot on a map, computer chips inside them perform an astonishing number of mathematical calculations. For a glimpse at the type of formulas involved, consider figure 5.2, which shows how Naval Research Laboratory mathematicians in 1971 formulated a solution for instantaneous navigation by satellites in four dimensions.

After locking onto a satellite signal the GPS receiver begins repeating the signal's digital navigation sequences the way people sing "Row, Row, Row Your Boat" in the round. It then delays its own sequences until they synchronize with the incoming signal. The amount of delay equals the signal's travel time from satellite to receiver, which, multiplied by the speed of light, yields the satellite's distance. With computed distances from three satellites and the accurate time from a fourth satellite's clock to synchronize the battery-powered quartz clock in the receiver, the receiver's computer can perform the geometric calculations (variously called trilateration, trilateralization, or triangulation) to determine a three-dimensional position as it moves in real time.

Society has become habituated to inexpensive portable computing power. Without it, satellite navigation would be unavailable to everyday users. Many technologies work together to make GPS perform in a way that would appear magical to prior generations. It is difficult to decide which one is most essential. But at the core of the system beat the "radio frequencies at the hearts of atoms"—the atomic clocks—so it seems especially fitting that the GPS constellation of satellites surrounding Earth resembles electrons orbiting an atom.

Going Public
The Roots of Civilian GPS Use

A little inaccuracy sometimes saves tons of explanation.

Saki (Hector Hugh Munro), "Clovis on the Alleged Romance of Business," The Square Egg, *1924*

Thirty-three thousand feet above the Sea of Japan, a Boeing 747 jumbo jet bored its way through the predawn darkness. In the cabin the lights were low, and most of the 240 passengers dozed. In the cockpit the pilot and copilot bantered with a sister airliner, also bound for Seoul, and hailed air traffic controllers in Tokyo to request permission to climb to thirty-five thousand feet. It was a routine procedure on the homestretch of a long flight. Having used most of its fuel, the lightened aircraft could now fly higher and faster.[1]

Minutes after completing the maneuver, the 747 was rocked by an explosion as an air-to-air missile blew a hole in the fuselage and severed the left wing. The enormous aircraft, which had mistakenly entered prohibited Soviet airspace, rolled to its left and began an uncontrolled spiral decent lasting more than twelve minutes before it crashed into the sea, killing all 240 passengers and 29 crewmembers.[2] The shoot down of Korean Air Lines Flight KE007 by the Soviet Union, in the early hours of September 1, 1983, became a milestone in Cold War superpower relations as well as a catalyst for the spread of GPS technology from military to civilian use.

Despite having triple-redundant inertial navigation systems (INS) on board, the pilot and copilot of Flight KE007 lacked what millions of motorists using GPS today take for granted—the positional awareness that comes from seeing a moving icon on a map. The pilots blindly trusted that the automatic pilot was faithfully executing, with INS guidance, the coordinates they had programmed into it.

Navigation methods have always relied on measuring a traveler's movement against a fixed reference point, such as the sun, a star, or the magnetic pole. INS uses gyroscopes to maintain that fixed reference point within the airplane itself, calculating latitude, longitude, and altitude electronically and feeding the information into the automatic pilot. At the time of the shoot down, commercial airplanes had been using INS for about a decade with no known simultaneous failure of all three systems.[3] The Flight KE007 incident led to speculation about how two experienced pilots could fly a civilian jetliner 360 miles off course over Sakhalin Island, home to one of the Soviet Union's most sensitive military installations. Explanations ranged from equipment malfunction to pilot error to conspiracy theories that the flight was on a covert spy mission. Numerous books and articles have examined the flight and these theories in detail, but it was not until 1993, after the breakup of the Soviet Union and a decade after the incident, that Russian Federation president Boris Yeltsin turned over the black boxes to investigators. The cockpit voice recording proved conclusively that the pilots were unaware they were off course (undoubtedly the reason Soviet military officials never acknowledged finding the black boxes and withheld the evidence). The most plausible explanation is that the pilots "armed" the INS but it never fully "engaged." This could have happened, because early in the flight the airplane was already several miles farther off course than the maximum distance the autopilot computer was programmed to accept when transitioning to automatic INS.[4] That left the airplane on a magnetic compass heading that carried it farther and farther off course. Having earlier sailed over Soviet territory above Kamchatka Peninsula and later failing to respond to radio contact on a frequency they were not monitoring or to notice bullets fired past the craft in the darkness, the pilots appeared to Soviet fighters to be executing an evasive maneuver with their routine "step-climb," a move that sealed the flight's fate.

A Calculated Response

In the days following the shoot down, President Ronald Reagan and his advi-sors drafted a National Security Decision Directive (NSDD 102) outlining the administration's response, which focused largely on strategies to marshal world opinion against the Soviets.[5] It is important to appreciate the dismal state of U.S.-Soviet relations at the time. Five months before, on March 8, Reagan had called the Soviet Union the "focus of evil in the modern world" in his famous "evil empire" speech.[6] On March 23 he delivered a prime-time address from the Oval Office announcing a program to develop a defense against ICBMs.[7] Officially named the Strategic Defense Initiative, or SDI, the program was tarred by critics with the moniker "Star Wars"—from the 1977 George Lucas film—because it proposed using exotic new technologies to destroy ICBMs during their flight through space.[8] By December 1983 the United States was set to begin fielding nuclear-tipped Gryphon cruise missiles and Pershing II intermediate-range ballistic missiles in Western Europe to counter the Soviet's SS-20 missiles.[9] Reagan's military buildup pushed federal spending in 1983 to its highest level as a percentage of gross domestic product since World War II.[10] Thus, it was within a general state of high anxiety that the Soviet regime committed the atrocity and responded to world reaction.

As administration actions under NSDD 102 gained traction, the U.S. Con-gress, the International Civil Aviation Association, the International Federa-tion of Air Line Pilot Associations, and the United Nations Security Council expressed condemnation through resolutions (with a Soviet veto blocking for-mal adoption of the UN resolution). Numerous airlines and entire nations sus-pended flights to or from the Soviet Union. For weeks, Reagan mentioned the incident in speeches and interviews, using the shoot down to draw a stark con-trast between Soviet values and behavior and those of democracies. Beyond apologies, Reagan pushed for reparations to the victims' families. There were passengers from thirteen countries aboard KE007, including sixty-six Ameri-cans.[11] One was Congressman Larry McDonald of Georgia, part of a six-person delegation en route to South Korea to mark the thirtieth anniversary of the U.S. defense treaty with that nation. The other five delegates were aboard Korean Air Lines Flight 015, which departed Anchorage only minutes after KE007 but was 350 miles away at the time of the shoot down.[12] Reagan also called for new aviation protocols to prevent a repeat of the tragedy in the future. Anticipating that twenty-four GPS satellites would be in orbit and operational by 1988, the

Reagan administration announced that it had decided to make the new navigation technology available to civil aviation.

Today, nearly every GPS historical timeline features Reagan's announcement, but most accounts either ignore the details or stretch the truth. Examples abound, particularly in poorly sourced online histories of GPS, of statements that Reagan "declassified GPS" or "gave it to private industry."[13] Very intelligent people have succumbed to such hyperbole. A high-profile example happened at the 2011 South by Southwest (SXSW) festival in Austin, Texas, during an onstage interview of Tim O'Reilly, the founder and CEO of O'Reilly Media, who is credited with coining the term *Web 2.0*. While making the point that people should view government as a platform that helps birth new technologies, O'Reilly dubbed Reagan "the father of Foursquare."[14] His remark drew an audible reaction from the audience and attracted lots of press coverage because of the unlikely coupling of the iconic president, who died in 2004, with the rapidly growing location-based service launched at SXSW in 2009. Foursquare employs GPS to allow its users, via smartphones, to share their location with friends, "check in" at restaurants, shops, and other hotspots, and take advantage of promotional offers targeted to them by businesses.[15] What neither the audience nor the press seemed to notice were the factual errors in O'Reilly's comments: "When the Navy and the Air Force put up the GPS system, they did not have to make the decision to open it up for civilian use. In fact, there was a lot of debate about that. It was Reagan, who after a *U.S. airliner* was shot down *over North Korea* because it strayed over *North Korean airspace*, said 'Hey, when you guys finish this GPS thingy, let's open it for civilian use.' It was an *executive order* that he gave" (emphasis added).

Recalculating the Facts

Reagan never issued any public executive order pertaining to GPS. (Some assert that NSDD 102, now largely declassified, contains references to GPS within portions that remain redacted.)[16] Furthermore, an Internet search of his speeches and interviews archived at the Ronald Reagan Presidential Library offers no evidence that he ever publicly uttered the phrase "Global Positioning System."

Reagan was vacationing in California when the shoot down occurred. His deputy press secretary, Larry M. Speakes, read a short statement on September 1 during a briefing with reporters at the Sheraton Santa Barbara Hotel. The next day, Reagan cut short his vacation, gave prepared remarks to reporters before boarding Air Force One for Washington, and issued Proclamation 5086, which ordered flags to be flown at half-staff at all federal facilities. After approving NSDD 102 on September 5, 1983, Reagan addressed the nation on television that night, playing a tape of the Soviet pilot communicating with ground control and saying (in Russian), "The target is destroyed." The president devoted his weekly radio address to the incident on September 17 and worked references into a speech—devoted almost entirely to the arms race—to the United Nations General Assembly on September 26.[17] On that very night, the world was closer to nuclear Armageddon than anyone knew. In an underground missile silo near Moscow, a computer screen began flashing an alarm that the United States had launched five ICBMs. Disaster was averted only because a Soviet lieutenant colonel, Stanislav Petrov, correctly concluded that the alarm was an error and chose not to react.[18]

In all of Reagan's public comments about the KE007 shoot down, and in numerous other settings in which he might have strayed onto the subject, he apparently never mentioned GPS. Speakes, it seems, made the only official public reference to GPS, while reading a prepared statement before a press conference September 16. It came at the end of the first paragraph in a brief, two-paragraph text: "World opinion is united in its determination that this awful tragedy must not be repeated. As a contribution to the achievement of this objective, the President has determined that the United States is prepared to make available to civilian aircraft the facilities of its Global Positioning System when it becomes operational in 1988. This system will provide civilian airliners three-dimensional positional information."[19]

That is all. There was no formal proclamation or presidential speech. (If NSDD 102 addresses civil aviation's use of GPS, it is unclear why information made public at a press conference would remain classified.) Given the public relations skills that earned Reagan the "great communicator" tag and the approach he used to announce SDI, it seems safe to say that if the GPS announcement were truly a change in policy, he would have opted to make a bigger splash. In any case, all GPS satellites orbiting at the time of Reagan's announcement already broadcast two signals—one for the military and another for civilian use.

In fact, from the earliest days of planning a navigation satellite system, government officials envisioned civilian use. When Deputy Secretary of Defense William P. Clements directed the services to create a joint program in April 1973, his memo included the following instructions: "The Joint Program Office would invite concerned non-DOD government agencies to participate in the

DNSDP, including program planning, user equipment design, and system tests. In addition, civil user needs should be considered in the design of the space-borne equipment."[20] Even before the Pentagon created the joint program in 1973, the top brass instructed NAVSEG, the interservice committee assembled to hammer out a joint navigation satellite system, to add air traffic control to the mix. In a November 2, 1970, memo to NAVSEG members, Chairman Harry Sonnemann wrote,

> The desire to consider how a navigation satellite system could satisfy the air traffic control needs has broadened the scope of the problem. The fact that the air traffic control portion probably should be active, and primarily oriented towards communications has decidedly slowed progress. The ever-tightening budget has also had a braking effect on the desire of the participants to be very positive in stating their needs or making any commitments. Thus, as I leave the chairmanship, I find that we are further away from a solution than when I took over, as the DOD systems concept has become one of the subsets of a national navigation problem.[21]

This single paragraph captures the dynamics surrounding early efforts to develop what would become GPS. Just as the Army, Air Force, and Navy had different needs and concerns, the interests of air traffic control were not the same as those of the military, which wanted a passive system—one that broadcast signals without any need for a user to activate it by transmitting a signal that would disclose his location.

Sticker Shock

Controlling costs—the original impetus for the Pentagon in seeking a single, multiuser navigation satellite system—remained an ever-present concern. Sonnemann recently recalled that after the Air Force system 621B was deemed too expensive given the number of users—fewer than one thousand aircraft that would be equipped with receivers—other users within the military services as well as from the civilian community were identified and solicited. The response from both quarters was tentative. "A wide range of military and potential civilian applications were identified, but those interested took a 'wait and see' position, that is, when the system's capabilities were verified with on-orbit data and its limitations identified, these potential customers would include GPS as another option to their existing capabilities, but not before," Sonnemann wrote.[22] This meant that despite its wide range of promising uses, getting GPS funded relied primarily on selling its ability to "drop five bombs in the same hole," a slogan that Parkinson, the first joint program manager, posted in his office.[23] Precision bombing capability was much on the minds of military leaders at the time, given their experience in the Vietnam War, when the United States conducted the most massive aerial bombardment in history. By weight, the bomb tonnage dropped in Indochina was three times as much as that used in the combined Pacific and European theaters of World War II and fifteen times the amount dropped in the Korean War.[24] Fewer bombs delivered more accurately held the promise of lower costs, fewer targeting errors, and reduced casualties among pilots and noncombatants.

Civil aviation remained on the sidelines during the GPS validation phase, which lasted five years. John McLucas, secretary of the Air Force when the GPS program began, left that post in 1975 to head the Federal Aviation Administration (FAA) for two years. Despite being such a strong proponent of the system that he ordered a vanity license plate that read "GPS NOW," he later admitted, "I could not get the FAA interested in GPS."[25]

The program faced continuing financial pressures as oil shocks, runaway inflation, and soaring interest rates pummeled the economy for the remainder of the decade. At the beginning of 1973, crude oil was around four dollars per barrel, the annual inflation rate was just over 3 percent, and the U.S. prime lending rate was 9.75 percent.[26] By December 1973, when the Defense Systems Acquisition Review Council approved GPS, the nation was two months into the Arab oil embargo, which lasted more than six months. Oil prices tripled, rising to twelve dollars per barrel, and 1973 ended with an annual inflation rate of 8.7 percent. The energy crisis was so severe that the White House Christmas tree remained unlit that year. If the GPS apps that people take for granted today had been available then, motorists would have been using them to search for the gas stations with the shortest lines or in many cases the ones that had any fuel at all to sell. Inflation reached 12.3 percent the following year before settling back into a 4–6 percent range that lasted until a second oil shock in 1979, following the Iranian Revolution. Crude oil then rose to twenty dollars per barrel on its way to a 1981 peak near thirty-five dollars, and inflation climbed to around 13 percent. By December 1980, the prime rate had soared to a record high of 21.5 percent.

In this tenuous budget environment, the GPS program gestated fitfully and nearly miscarried. As a support system rather than a weapons system, it lacked dependable support from one or more service branches—even from the Air

Force, which served as executive agent of the joint program. When President Jimmy Carter canceled the B-1 bomber in 1977, the Strategic Air Command dropped plans to acquire six hundred GPS receivers. The Air Force then postponed satellite purchases and delayed launches.[27] Whereas the original plan would have yielded limited operational capability at the end of the test phase, the decision to delay pushed any practical use of GPS further into the future.

In January 1979, one month before a scheduled second DSARC hearing to review test data prior to authorizing full-scale engineering development and production, the General Accounting Office (GAO; renamed the Government Accountability Office in 2004), issued a blistering report to Congress. Its title, *The NAVSTAR Global Positioning System: A Program with Many Uncertainties*, was more reserved than its findings. The validation studies, originally scheduled to run from December 1973 to March 1978, had slipped fourteen months by early 1979, and their cost had risen from $178 million to more than $406 million. Together with full engineering and production costs, the total program estimate had more than doubled, to $1.7 billion, $900 million more than originally budgeted.[28] Moreover, GAO found that the program budget did not include $2.5 billion in related future costs for user equipment, satellite replenishment, and space shuttle launches. That raised the projected total program cost to more than $4.25 billion, the report estimated. Beyond schedule slippages and budget overruns, GAO criticized the program for failing to justify its high cost with hard data on the number of users or by identifying cost savings that would come from replacing existing systems. By this time, Bradford Parkinson had retired from the Air Force and gone to work for Rockwell International, the contractor building the GPS satellites. Col. Donald W. Henderson succeeded him, and the deputy program manager, Col. Steve Gilbert, moved to a post at the Pentagon, where he advocated for GPS development.[29]

GPS survived the second DSARC hearing (finally held in June 1979), which authorized full-scale production, but just six months later, in December 1979, the Pentagon made across-the-board budget cuts of $512 million, or roughly 30 percent of defense spending, for fiscal years 1981 to 1986. This led the Air Force to scale back the constellation from twenty-four to eighteen satellites—too few to achieve the promised accuracy—and for three years, the Air Force "zeroed out" the program, effectively mothballing it.[30] In June 1980, for example, the Air Force requested $16.3 million, a mere 6 percent of the $234.5 million the joint program office had requested.[31] Whether these cuts represented simply a lack of support or a canny maneuver to shift more of the financial burden for the multiservice program outside the Air Force is unclear. Either way, backing from such top leaders as Assistant Secretary Donald Latham in the Office of the Secretary of Defense saved the program, as that office reinstated funding each year.[32]

The Air Force cuts did not help win support in Congress. During the fiscal year 1982 defense budget authorization process, the House Armed Services Committee recommended terminating the program. Some observers have attributed this largely to the loss of the powerful committee member Rep. Charles H. Wilson, a Democrat from California, whose district included the GPS contractor Rockwell International and other defense plants.[33] In 1980 Wilson lost the seat he had held since 1962 after the House censured him for lying about cash gifts he accepted from a foreign operative who was trying to influence a decision about pulling U.S. troops out of South Korea. Without Wilson advocating for the program, the argument goes, committee members adopted the narrow framework GAO used to evaluate costs and benefits—merely identifying existing users who would switch to GPS without trying to assess ways the system might transform military doctrine.[34] Visionary thinking may not have been the strong suit of GAO, but the Senate was more receptive to the possibilities, issuing a remarkable (given the sums of money required) "build-it-and they-will come" endorsement. "It may be difficult to understand the full potential until the system is deployed and the vast number of potential users are able to see what it will do for them," stated the fiscal year 1981 Senate Authorization Report.[35] With President Reagan urging a massive defense buildup, the Senate view prevailed.

Another factor probably influenced many lawmakers. Terminating GPS would have also killed a less publicized secondary use of the satellites, distinct from its navigation mission but tied to its precise positioning capability—the Integrated Operational Nuclear (Detonation) Detection System (IONDS).[36] Arms control verification relies on the ability to monitor nuclear weapons tests. The United States first fielded ground sensors to detect nuclear detonations in the late 1940s, and it launched space sensors aboard satellites in the mid-1960s.[37] These devices detect nuclear bursts using instruments highly sensitive to proton, electron, neutron, x-ray, and gamma radiation. Under the Vela (short for *velador*, a Spanish term for a watchman or guard) Program, the Air Force began launching satellites into high orbits—about a fifth of the way to the moon—six days after the United States, Great Britain, and the Soviet Union signed the Limited Test Ban Treaty on August 5, 1963.[38] By 1970 the Air Force had launched a dozen Vela satellites, but the satellite's design life span was just eighteen months, so the need for frequent replacement was great. (Although one satellite operated for fourteen years.)

Soon after the GPS program began, military officials investigated the possibility of "piggybacking" nuclear detonation sensors aboard GPS satellites. In tests conducted in 1978 and made public in 1982, satellite contractor Rockwell International studied whether it could add IONDS sensors to GPS without negatively affecting the primary navigation mission. Rockwell concluded that GPS was an "ideal host" for nuclear detonation surveillance.[39] Worldwide coverage and the ability to pinpoint the location and altitude of a nuclear burst made GPS ideally suited for the job. Spreading the sensors across eighteen satellites (the planned constellation in 1981) made them more likely to survive any Soviet antisatellite attack, a Congressional Budget Office (CBO) report noted.[40] The CBO report further stated that in a nuclear war IONDS could tell U.S. commanders which areas of the United States had escaped destruction, helping to coordinate recovery efforts, as well as identify Soviet targets that had escaped an initial retaliatory strike, aiding decisions about subsequent strikes. It is not difficult, given the mindset of the Cold War, to conclude that for some officials these capabilities outweighed GPS's still-unrealized navigational potential. The CBO report, citing national security reasons, withheld budget figures for IONDS. However, the Department of Defense's annual report for fiscal year 1981 openly listed projected costs of $40.5 million for IONDS development in fiscal years 1979 through 1982.[41] NAVSTAR 6, launched April 26, 1980, carried the first IONDS sensors, and after successful testing, every subsequent GPS satellite has carried nuclear detonation sensors.

Escalating costs for the GPS program continued to be a concern, as a GAO report issued in February 1980 demonstrates. It offered a new estimate of $8.6 billion in total program costs through the year 2000, offset by no more than $1.2 billion in savings identified from the phaseout of other defense systems. On the other hand, the report stated, "We believe that force-effectiveness studies have demonstrated that NAVSTAR could improve the effectiveness of some military missions."[42] Beyond reducing the number of aircraft needed to achieve a particular objective and delivering munitions more precisely, the report listed en-route navigation, search-and-rescue operations, and minesweeping. The report also noted that the Army, Navy, Air Force, and Defense Mapping Agency had committed to purchase 14,828 receivers over sixteen years, starting in 1984. If these projections for program cost and number of committed users had remained static through 2000, the military would have spent about a half-million dollars per receiver to provide GPS service. Of course, as the system reached fruition over the next two decades the number of military users rose significantly, and the number of commercial users surpassed those in the military.

Down to Earth

The tiny portion of those first receivers allocated to the Defense Mapping Agency—fifty in all, or less than 1 percent—was disproportionate to their significance. Indeed, land surveying emerged as the first application of GPS to cross over to civilian use. While there were too few satellites in orbit in the early 1980s for practical navigational use, surveyors did not require real-time calculations. They could record observations using signals whenever the satellites passed overhead and process the data later. GPS saved time and boosted productivity so much that it cost about 10–20 percent of the cost of conventional surveying while offering accuracy three times that of existing methods.[43] The National Oceanic and Atmospheric Administration (NOAA), which is part of the Department of Commerce and oversees the National Geodetic Survey, published the first technical standards for civil GPS use in the *Federal Register* in 1984. This step helped convince the surveying industry that GPS signals— generated by a military system—would be available for commercial use.[44]

Surveyors for years had been using inertial surveying systems (ISS), which employed gyroscopes and the same principles as the INS systems used on airliners, as well as Doppler systems that relied on radio signals from Transit (the Naval Navigation Satellite System, or NNSS). This use of Transit traces its lineage directly back to the Vanguard satellite proposal, which envisioned the use of satellites for geodetic purposes. The earliest GPS receivers, designed in the 1970s to test the fledgling fleet, were built under defense contracts, but electronics manufacturers already had been building survey instruments using radio signals. As the GPS constellation grew, it was natural for some to see a new market for surveying use. The first commercial GPS receivers appeared around 1982, including the STI-5010, built by Stanford Technologies (similar to those developed for the military's GPS ground tracking stations); the Macrometer V-1000, designed by Massachusetts Institute of Technology researchers and marketed by Litton Aero Service; and the Texas Instruments TI 4100, known as the "NAVSTAR Navigator."[45]

At the time, six of twelve satellites planned for the first block of the GPS constellation were in orbit—the first four launched in 1978 and two more in 1980—all lifted into space by Atlas F rockets from Vandenberg Air Force Base in California. NAVSTAR 7 would have been available, but the launch on December 18, 1981, failed. These six satellites provided about four hours of useable coverage daily over the United States, but their orbital configuration brought them into view four minutes earlier each day, making GPS signals available predominantly during daylight hours for six months of the year and predominantly at night for the other six months.[46]

The Macrometer was nothing like the handheld or pocket-sized devices common today. It was a cubical metal box roughly twenty-five inches on each side, weighing about 160 pounds, with a separate antenna that weighed another forty pounds. Equally hefty was the price—$250,000 in 1982 dollars.[47] Transporting these units to a survey site often required tethering them beneath a helicopter.[48] Users stored the field data they collected in small cartridges and processed the information later. Because technical limitations prevented the use of GPS signals for clock synchronization, surveyors needed two of these behemoths, synchronized to each other, making the cost and effort suitable only for larger projects.[49]

While the Macrometer sounds massive by today's standards, it was svelte compared to the first GPS receiving unit Rockwell Collins built for the Air Force to test the fledgling constellation in 1977. The Generalized Development Model (GDM) was the size of a tall bookcase, weighed 270 pounds, and stood with two high-backed seats for its operators atop a large, square pallet on casters, designed for loading onto aircraft during flight tests.[50] The TI-4100, by comparison, was a trim fifty-three pounds, about the size and appearance of a small microwave oven, with a separate six-inch, cone-shaped antenna and a handheld keypad and display screen connected by a coiled cord.[51] This receiver, which could simultaneously track four GPS satellites, marked a transition toward more, and more practical, commercial applications.

In March 1985, with nine GPS satellites in orbit, Texas Instruments introduced a new software package called Satplan, which allowed users to create a precise satellite availability schedule in tabular form by date and location. The company's announcement, which appeared in such publications as *Maritime Reporter and Engineering News*, called it "an enhancement that makes the Global Positioning System (GPS) more productive as a navigation/positioning tool."[52] The software ran not on the receiver itself but on the TI Portable Professional Computer (PPC). A quick review of its features illustrates how far both GPS

Fig. 6.1. Generalized development model by Rockwell Collins. (Courtesy Rockwell Collins)

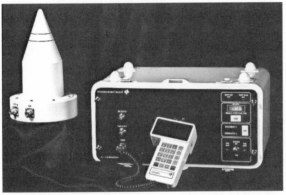

Fig. 6.2. TI 4100 and antenna by Texas Instruments. (Courtesy Texas Instruments)

receivers and computing have come. The TI PPC offered 256 kilobytes of random access memory (RAM) and a first-generation Intel 8087 math coprocessor, designed to boost the 8086 central processing unit (CPU), which had a clock-speed rating of five megahertz (five million cycles per second).[53] This was the CPU that begat computers bearing the 286, 386, 486, and finally Pentium trademarks. At more than twenty-six pounds, this box and keyboard combo with a nine-inch screen was far from the portability of a laptop computer. Uploading programs and storing information depended on five-and-a-quarter-inch floppy disks. The list price was about $3,000. By comparison, the most expensive smartphones today cost no more than $700, typically have 512 megabytes of RAM—about two thousand times as much computing space—and have CPUs running a billion cycles per second. They wirelessly connect with the Internet to retrieve and display tabular charts like those Satplan created—all while a GPS receiver built on an integrated semiconductor chip small enough to fit into a wristwatch runs unobtrusively in the background, providing continuous data to multiple location-based applications running simultaneously.

Such connectedness was undreamed of in the mid-1980s, but the advance of semiconductor technology and miniaturization of electronic devices was underway. After the TI-4100, the size and cost of GPS receivers began the familiar downward curves associated with most electronic products, but a delay in completing the constellation itself slowed the expansion of GPS applications to other uses, particularly navigation.

A Major Malfunction

The February 1980 GAO report warned about a potential problem that could knock the GPS program off schedule and did so in such a deadpan manner that reading it today is rather chilling: "Space Shuttle problems could also jeopardize DOD's plan to have NAVSTAR fully operational by 1987."[54] The catastrophic failure of the space shuttle *Challenger* seventy-three seconds into its flight on January 28, 1986, undoubtedly exceeded the concerns GAO had in mind. It was the twenty-fifth shuttle mission, officially designated STS-51-L. Although rocket launches routinely experienced technical problems and delays, shuttle missions had become routine—considered safe enough to add schoolteacher Christa McAuliffe to the *Challenger* crew. Starting with two launches in 1981, NASA successfully launched three missions in 1982, four in 1983, five in 1984, and nine in 1985. Only ten days before the disaster, on January 18, *Columbia* (which later disintegrated on reentry on February 1, 2003) had completed a six-day mission.[55]

The Air Force had planned to launch two dozen Block II satellites aboard the shuttle by 1987. Recall, however, that funding cuts in 1980 and 1981 trimmed the constellation size to eighteen and added a year to the schedule. The eighth through eleventh Block I satellites, not designed for the shuttle, flew aboard Atlas rockets from 1983 to 1985. A twelfth Block I satellite, which was removed from the launch schedule and converted into a prototype Block II satellite for test purposes, resides today at the San Diego Air & Space Museum and is the only GPS satellite on public display.[56] The grounding of shuttle flights after the *Challenger* disaster slowed the GPS launch schedule by about two years as the Air Force developed plans to use a new "medium launch vehicle" called Delta II for the first eight Block II satellites.[57] Although it was a new design at the time, the Delta II rocket, as well as launch pads 17A and 17B at Cape Canaveral, traced their lineage directly to the Air Force's THOR intermediate-range ballistic missile program of 1956.[58] Over several decades, the Delta rocket, with subsequent upgrades, has been used extensively for both military and commercial payloads. Between February 14, 1989, and August 17, 2009, Delta II rockets carried forty-eight GPS Block II satellites into orbit.

The second block of GPS satellites consists of four generations or groups of updated and revised designs, with the following designations:[59]

IIA ("A" for advanced)—Nineteen satellites developed by Rockwell International were launched between November 1990 and November 1997. Although designed to last seven and a half years in space, two have operated more than twenty years.

IIR ("R" for replenishment)—Thirteen satellites built by Lockheed Martin were launched between July 1997 and November 2004 to replace previous Block II and IIA satellites that had outlived their design life span. A key improvement in these satellites was onboard clock monitoring, which boosted accuracy and allowed longer intervals between ground station updates.

IIR(M) ("M" for modernized)—Eight satellites developed by Lockheed Martin were launched between September 2005 and August 2011. This design added a second civilian signal (L2C) for better commercial performance and two new, more powerful military signals with enhanced jam resistance.

IIF ("F" now stands for follow-on; earlier references used the word *future*)— The first of twelve IIF satellites, developed by Boeing (which acquired Rockwell International's aerospace and defense businesses in 1996), launched on May 27, 2010, with subsequent launches scheduled through 2013. These satellites boast a twelve-year life expectancy, more accurate atomic clocks, and a third civilian signal (L5) for improved transportation safety, especially aviation. (A new generation of satellites, Block III, will begin launching in 2015. See chapter 10.)

During the delay in GPS satellite launches resulting from the *Challenger* disaster, surveying equipment sales kept the commercial market growing and fueled research and development of new GPS applications at a faster rate than would have occurred through defense contracts alone.[60] To address problems in the field, advanced techniques and equipment evolved, such as the development of complex interconnected receivers that correct for errors by calculating the

Fig. 6.3. GPS Block IIA satellite. (Courtesy National Coordination Office for Position, Navigation and Timing)

differences in signals received at one or more (static) base stations, or precisely known, surveyed positions, with signals from receivers at other, unknown positions, including moving (kinematic) receivers. The development of "differential GPS" greatly enhanced accuracy and led the way toward land-based augmentation systems that improve navigational accuracy around marine ports and airports.

Smaller, Lighter, and Faster

In 1988 three companies introduced GPS receivers small enough and light enough to hold in one hand. Rockwell Collins, a defense contractor for both GPS satellites and receivers, demonstrated a prototype of the first compact, all-digital receiver, dubbed "Virginia Slim" after a popular cigarette brand

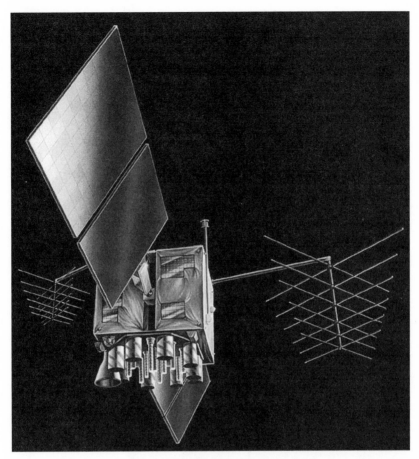

Fig. 6.4. GPS Block IIR satellite. (Courtesy National Coordination Office for Position, Navigation and Timing)

because it was about the size of a pack of one hundred millimeter, or king-size, cigarettes.[61] Magellan followed with the first commercial handheld GPS receiver, the NAV 1000, which resembled a walkie-talkie. The device measured about seven and a half inches by three and a half inches, weighed less than two pounds, and could track four satellites by automatically sequencing their signals on a single channel.[62] Trimble Navigation offered the Trimpack, a three-channel sequencing receiver capable of tracking eight satellites.[63] Trimpack had a slightly larger but flatter design than the NAV 1000. Measuring roughly nine inches wide by two and a half inches tall by eight inches deep and weighing just over three pounds, it could be held by hand or bracket-mounted on a

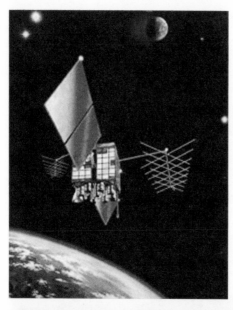

Fig. 6.5. GPS Block IIR(M) satellite. (Courtesy National Coordination Office for Position, Navigation and Timing)

Fig. 6.6. GPS Block IIF satellite. (Courtesy National Coordination Office for Position, Navigation and Timing)

vehicle dashboard or boat instrument panel. Both of these commercial receivers set the stage for broader use of GPS for navigation.

With public use of GPS growing and inquiries about it proliferating, the Pentagon asked the Department of Transportation in 1987 to create an office to coordinate civilian use. In 1989 the U.S. Coast Guard became the official liaison between military and civil users through creation of the Navigation Cen-

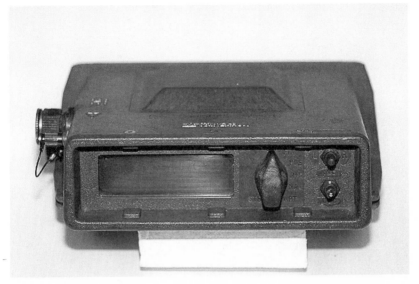

Fig. 6.7. Trimpack GPS receiver by Trimble. (Courtesy U.S. Army Heritage Museum)

ter, or NAVCEN, located in Alexandria, Virginia. That year the GPS program successfully launched five Block II satellites into orbit.

As the 1980s ended, fifteen GPS satellites were in orbit, but the constellation was several years from being complete and declared operational by the Pentagon. Civilian GPS use was increasing, although it was confined largely to specialized tasks due to the size and cost of receivers and the gaps remaining in coverage. The terms *Global Positioning System* and GPS remained far from household words. It would not be long, however, before GPS got a highly visible public demonstration—a debut with fireworks: the Persian Gulf War.

Going to War

GPS Aids Military Success in the Persian Gulf

7

A line has been drawn in the sand.

George H. W. Bush, fifty-fifth presidential news conference,
White House briefing room, August 8, 1990

As day broke on January 16, 1991, seven B-52 bombers rumbled more than nine thousand feet down the runway at Barksdale Air Force Base in Louisiana, finally lifting into rainy skies and heading for targets more distant than any bombing run previously ordered.[1] After five months of military preparations and diplomatic maneuvering following Saddam Hussein's August 2, 1990, invasion of Kuwait, Operation Desert Shield was transitioning to Operation Desert Storm.

Not only clouds but also extreme secrecy shrouded the mission. Designated officially as Operation Senior Surprise, the attack from Barksdale carried such a top-secret classification that the several-dozen participants did not even utter its code name. They dubbed the mission Operation Secret Squirrel, a nickname (taken from a cartoon character) for the weapons they carried, which officially did not exist.[2] Fifty-seven aviators conducted the mission, with each plane's crew including two sets of pilots and navigators who exchanged shifts during the nonstop thirty-five-hour round trip. After flying seven thousand miles in fourteen and a half hours, refueling twice along the way, the planes reached designated positions over northwestern Saudi Arabia, where their crews initiated the first combat use of new GPS-guided cruise missiles that had been in development since 1986. The planes carried thirty-nine missiles—most of the existing stock remaining after tests consumed eleven of the original fifty-seven missiles the Air Force had ordered.[3] Four of the missiles on the planes experienced software problems and were not launched. The remaining thirty-five missiles were launched at eight targets across Iraq, from Mosul in the north to Basra in the south. The target list, comprising power plants, electric transmission facilities, and military communications hubs, reflected a strategy of swiftly knocking out Iraq's air defense system and rendering Iraqi command and control "deaf, dumb and blind."[4]

The flight home proved more difficult than the attack phase. Struggling with bad weather, stronger than expected headwinds, engine and radio problems, and the drag of unlaunched missiles, the B-52 pilots had to refuel more often, and the return to Barksdale took more than twenty hours. When they finally touched down the pilots hurriedly taxied the planes into their hangars to keep the unused weapons out of sight.[5] This historic combat mission, the first one launched in Operation Desert Storm (though not the first to fire weapons, as attack aircraft already in theater began bombing two hours before the B-52s arrived), was the one the public would learn about last in a war that unfolded with unprecedented visibility on the CNN cable news channel. Operation Senior Surprise remained secret until January 16, 1992, exactly one year later.[6]

GPS Enters Guided Weaponry

The extra secrecy before and after the mission arose from the missile's origin and warhead, in addition to its guidance system. Development began after the Air Force accidentally hit the French embassy in Libya with an errant bomb and lost an F-111 fighter and its crew during Operation El Dorado Canyon, an April 1986 raid on Libyan military facilities and terrorist training camps. The Air Force sought a precision weapon that its pilots could launch from safer distances and chose to modify its only standoff weapon, the nuclear-tipped AGM-86B air-launched cruise missile (ALCM). Bomb makers replaced the AGM-86B's nuclear warhead with a one-thousand-pound conventional bomb and adapted its inertial navigation system for guidance using a GPS receiver in place of the existing terrain contour matching (TERCOM) system. TERCOM uses onboard altitude-sensing radar to compare the terrain beneath the flying missile to topographic maps of the intended flight path stored in its computer.[7] Programming each flight is time consuming, and each missile launch must occur very close to its programmed start point, so the system can lock in on the terrain. This works best where the terrain is bumpy and unique. Iraq's flat desert forced mission planners to send weapons using TERCOM zigzagging among whatever landmarks were available.

GPS, by contrast, provides easier targeting and more launch flexibility. If the missile has precise target coordinates and the receiver can acquire signals, it can determine its location and the route to its destination in the same way a dashboard navigation system will respond by "recalculating" when the driver veers off the expected path. That sounds commonplace today, but in the mid-1980s there was no precedent for having a single-channel GPS receiver in a missile acquire multiple satellite signals on the fly, much less for having it feed that information seamlessly into an inertial guidance system.[8] Boeing engineers overcame those and other technical challenges within twelve months.[9] The resulting weapon was the AGM-86C conventional air-launched cruise missile, or CALCM. The missile is nearly twenty-one feet long, two feet in diameter, and has a wingspan of twelve feet. Weighing more than three thousand pounds and powered by a turbofan engine, it can cruise up to 690 miles (its unclassified range) at speeds of 500 mph.[10]

Anyone knowledgeable about ALCMs who saw the B-52s en route to the Persian Gulf might have reasonably suspected the imminent use of nuclear weapons, since the secret conventional missile looked just like the nuclear version. Extreme secrecy was necessary not only regarding the mission, which gave the Air Force a potent surprise against the adversary, but also regarding the development of the CALCM missiles. Had news of the existence of GPS-guided cruise missiles leaked out, the Iraqi military could have predicted possible attack times. With the GPS constellation incomplete and only sixteen satellites in orbit, there were gaps in coverage, leaving specific windows of opportunity when the required minimum four satellites would be overhead for several hours. Mission planners drew the flight timetable carefully to avoid any possibility that Libyan radar operators might spot the B-52s over the Mediterranean Sea and warn Iraq before the first F-117 stealth fighters attacked.[11] Disclosure of the CALCM's existence could also have complicated negotiations in the Strategy Arms Reduction Treaty (START I), signed July 31, 1991, by President George H. W. Bush and USSR president Mikhail Gorbachev.

Fig. 7.1. AGM-86C/D conventional air-launched cruise missile. The first GPS-guided missiles used by the Air Force were modified AGM-86B cruise missiles with conventional bombs replacing nuclear weapons, but they looked identical. (USAF photograph)

Only seven other GPS-guided missiles were used in the Persian Gulf War— all of them Navy AGM-84 standoff land attack missiles, or SLAMs. That weapon's development flowed from circumstances similar to those the Air Force experienced in Libya. The Navy sought to put more distance between its pilots and surface-to-air missiles after losing aircraft in raids on Syrian-backed forces in Lebanon during the 1980s. When an A-6 Intruder and an A-7 Corsair were shot down in 1983, the Navy began modifying the twenty-year-old A-6 aircraft and created a new training program for strike pilots.[12] After two more A-6 aircraft were lost in 1986, the Navy began developing SLAMs.[13] Like the Air Force, the Navy did not start from scratch. The SLAM combined the airframe, turbo-jet propulsion system, and five-hundred-pound warhead of the Navy's antiship Harpoon missile with a GPS-aided inertial navigation system and targeting technologies borrowed from other missiles. Such targeting technologies included an infrared sensor that transmitted video to the cockpit of an accompanying plane and an electronic data link that allowed the plane's weapons officer the ability to control the missile's aim point using the video image.[14] The nearly fifteen-foot, 1,400-pound missile could fly slightly more than six hundred miles per hour to hit targets up to sixty-eight miles away. (Newer versions have extended the range to 143 miles.)[15] The SLAM was the Navy's newest weapon when Iraq invaded Kuwait in August 1990; it had never been fired in combat, and fewer than fifty test missiles were available.[16] A-6 Intruders

Fig. 7.2. AGM-84 stand-off land attack missile. The Navy's first GPS-guided missile, the AGM-84 was a modified Harpoon antiship missile, later also deployed from aircraft. (U.S. Navy photograph)

from the USS *John F. Kennedy* in the Red Sea fired two SLAMs at a uranium enrichment plant on the second day of the war, January 18, and fired a third at the same target a week later, on January 25. Aircraft from the USS *Saratoga*, also in the Red Sea, fired four SLAMs—two from an A-6 Intruder at a Kuwaiti radio tower the Iraqis used for military communications, and two others from an F/A-18 jet fighter at a military building at Taji Airfield, about twenty miles north of Baghdad. Based on the video images these weapons transmitted, four of the seven hit their targets; assessing the damage caused is difficult since the video ceased on impact. Difficulty with the electronic data link and/or loss of GPS signals affected the three missiles that did not perform as expected.[17]

Strikes Yield Cautious Assessments

Four hits in seven tries is not a horrible ratio for a weapon rushed into service, but too few missiles were launched to draw statistically meaningful conclusions. That was apparently also true of the CALCMs. The thirty-five missiles the Air Force fired were later *reported* to have achieved 85–91 percent of the mission's objectives, with only two going astray, as many as thirty-one hitting their targets squarely, and one slicing its aim point—a telephone pole—in half. Hard numbers with official endorsement are elusive.[18] Queried by reporters about the missile's accuracy when the weapon was revealed during a press briefing, Pentagon spokesman Pete Williams said, "The Air Force considers it to have been very effective. But the experts who analyzed this don't believe that it is absolutely possible to pinpoint the precise success rate."[19] The Pentagon's *Conduct of the Persian Gulf War: Final Report to Congress*, delivered in 1992, states that the CALCM (which at the time was still called an ALCM) "played an important role in Operation Desert Storm." It provides no statistics but instead offers this cautious appraisal: "A complete assessment of the AGM-86C's effectiveness is difficult to determine because of incomplete battle damage assessment (BDA) and the inability to distinguish damage caused by other munitions that struck some of the same targets. All missiles launched successfully transitioned to cruise flight. Demonstrated accuracy appears consistent with the results obtained from testing."[20]

In other words, to ensure success the military intentionally struck many targets repeatedly using a variety of bombs. The use of new GPS-guided missiles that were untested in battle would certainly have warranted such a strategy. A major review, the *Gulf War Air Power Survey*, released in 1993 by an independent, Air Force–appointed panel, was similarly cautious: "The CALCM's high explosive fragmentation warhead is designed to attack soft targets. Nevertheless, CALCM was apparently effective in Desert Storm against electrical generator switching facilities and exposed communications relay facilities. In contrast to TLAM [Tomahawk land attack missile], generalizations concerning CALCM effectiveness in Desert Storm must be treated with caution in light of the small number fired."[21]

In any case, the thirty-five CALCMs were sufficient to impress the Air Force to order hundreds more, and today the missiles feature upgraded avionics with multichannel GPS receivers and three-thousand-pound bombs.[22]

The TLAM referenced in the survey is a TERCOM-guided, deep-strike weapon the Navy launched from ships and submarines against heavily defended Iraqi targets as far as seven hundred miles away. (It now has a listed maximum range of one thousand miles and uses GPS.)[23] The Navy fired 288 TLAMs during the Persian Gulf War—116 during the first twenty-four hours and nearly two-thirds of the total within the first forty-eight hours of the six-week air campaign.[24] All but six achieved cruise flight. Footage of these twenty-foot-long missiles riding fiery plumes from the decks of destroyers, sometimes against a pitch-black sky or breaking the surface after a submarine launch, were among the most vivid images televised during the war. They form a sort of visual bookend with grainy, black-and-white videos of laser-guided missiles (different weapons, not TLAMs) perfectly aligned in the crosshairs diving into airshafts on rooftops. These images provided a televised, round-the-clock public introduction to precision-guided weapons, or "smart bombs," which had been under development since the Vietnam War. However, war coverage did not produce an equal public introduction to GPS for a variety of reasons—

the Air Force CALCMs remained secret, the Navy used so few SLAMs, receivers were in short supply, the ground war was brief, and the complex system was not easily explained.

Call It What You Will

As the military pressed GPS technology into service ahead of schedule for the war, the news media faced the challenge of introducing unfamiliar terminology to an audience of millions. The approaches varied and the results were uneven. Prior to Desert Storm, detailed reporting about GPS remained mostly the province of professional journals like *Aviation Week & Space Technology*, the *Journal of Surveying Engineering*, *Mining Magazine*, or *Offshore* (aimed at the oil and gas industry). *GPS World*, a monthly publication devoted to the technology, started in January 1990. Business-oriented publications such as the *Wall Street Journal* and *Forbes* magazine wrote about GPS from its formative years primarily because the defense contractors building the system were important to investors. Stories for more general audiences were sporadic. An early example appeared in the sports section of the *Palm Beach Post* on November 5, 1989, following the Palm Beach International Boat Show. "Global Positioning System Will Be Rave of the Future" read the headline. In the story vendors and boating enthusiasts commented on the buzz created at the show by GPS receivers, including Magellan Corporation's new handheld model, which cost $2,995, and Trimble Navigation's $9,000 marine unit. The article included a good explanation of how GPS works.

As the troop buildup in Saudi Arabia progressed during the fall of 1990, the *CBS Evening News* aired a segment with Scott Pelley interviewing Marines. He referenced a network of satellites beaming navigation instructions to troops in the desert and identified it, saying, "The network is called the global positioning system."[25] However, the focus was not on the system's capabilities but rather on the shortage of receivers. Four days before the air campaign began *ABC Weekend News Sunday* concluded its broadcast with a report from the Winter Consumer Electronics Show in Las Vegas, where Pioneer introduced its Avic One satellite navigator for automobiles.[26] Reporter Tom Schell explained that it used the "global positioning satellite network" to update the car's location to within forty feet. The program did not mention the satellite system's military significance or connection to the impending war.

After the shooting started most war coverage lacked the time and space to provide many details about GPS. In much the same way that reporters in the

1950s introduced satellites to a general audience by calling them artificial moons, reporters covering Desert Storm often commingled GPS with reconnaissance, communications, and weather satellites. For example, on the February 8, 1991, CBS Evening News, Bruce Hall reported on the role of spy satellites and noted that the military had "a group of satellites" that could tell ground troops their location on the featureless desert. An unidentified soldier he interviewed remarked simply that the system worked great.[27] Some reports artfully depicted the system's utility. In a February 25, 1991, Los Angeles Times article, "Ground War Puts Some Exotic New Weapons Systems to Test," staff writers Karen Tumulty and Bob Drogin described the unnamed system as "a two-pound, high-tech compass that gets its readings from satellites rather than magnetic poles."[28] A more in-depth wartime look at GPS was a February 6, 1991, special to the New York Times business section, titled "Business Technology: War Spurs Navigation by Satellite."[29] Andrew Pollack's story fully explained the technology, its military and emerging civilian uses, and the rise of receiver manufacturers Trimble Navigation and Magellan Systems. Some war coverage billed as an examination of high-tech military systems overlooked GPS altogether. CNN's Crossfire devoted the entire program on January 25, 1991, to the issue of whether opponents of Reagan's military buildup owed an apology given the success of American high-tech weapons. GPS was not mentioned in a discussion that included aircraft carriers, stealth fighters, cruise missiles, smart bombs, Patriot antiballistic missiles, the M-1 tank, and the Bradley fighting vehicle.[30] A CBS News special report about fighting a high-tech ground war, which aired February 23, 1991, mentioned multiple-launch rocket systems, the night vision and laser-targeting capabilities of the Apache helicopter, and the speed and accuracy of the M-1 tank, but not GPS.[31]

The war's end brought countless media analyses and spawned many books. Turner Broadcasting, CNN's parent company at the time, rushed a book to market in the summer of 1991. It of course mentions nothing about the classified CALCM attack. Numerous references to "smart bombs" explain bombs and missiles guided by lasers, TV cameras, or infrared "heat-seeking" systems. Its index lacks the terms NAVSTAR, Global Positioning System, or GPS, and none of the three appear explicitly in the book, although a few passages come close to mentioning them. One illustration describing the SLAM states, "While the missile is in flight, the *satellite receiver/processor* updates the missile's inertial navigation system" (emphasis added).[32] In another instance the text states, "Also overflying the Middle East were U.S. global-positioning, meteorological, and launch-warning satellites. Positioning satellites permitted U.S. troops on the ground to use hand-held receiving devices to determine their exact position without reference to any ground features."[33] These wording choices suggest careful editorial decisions recognizing the public's lack of familiarity with GPS. Two years later, a comprehensive and popular account of the war, *Crusade: The Untold Story of the Persian Gulf War*, by the Pulitzer Prize–winning journalist Rick Atkinson, avoided the terms NAVSTAR, *Global Positioning System*, or GPS. Its account of the by-then declassified CALCM mission discusses the arms control issues posed by replacing the nuclear warheads but omits any mention of GPS guidance.[34]

Despite its complexity and the varied ways newspapers, television, and books explained or worked around it, GPS emerged from the war with much broader name recognition. One other medium deserves mention—word of mouth. More than 540,000 coalition troops from thirty-one countries discovered a new battlefield dynamic as their units navigated aircraft to precise weapons launch points or crossed vast expanses of featureless terrain in vehicles or on foot. These troops witnessed the utility of GPS and returned home after the war to talk about their experiences. Many were probably the earliest adopters of GPS navigation devices over the following decade.

Consider this passage from the *Gulf War Air Power Survey*, describing how GPS improved the delivery of conventional weapons: "Although laser-guided munitions constituted only 6.7 percent of bombs dropped from tactical aircraft during Desert Storm, accurate bombing played a pivotal role in the exercise of air power by Coalition and particularly U.S. air forces. The relatively low percentage of precision-guided bombs reflects in part the fact that many of the unguided bombs were dropped from 'smart' platforms (e.g., aircraft) that were, at least in principle, capable of achieving near precision-guided munitions accuracy with 'dumb' bombs."[35]

Despite media fascination with "smart" bombs, roughly 210,000 bombs used during Desert Storm were unguided.[36] The U.S. military had installed GPS equipment in only about 300 aircraft before the war and sent some to the theater outfitted with portable receivers.[37] About half of the 66 B-52 bombers, less than a third of the roughly 250 F-16 fighters, and a handful of the 82 F-111 fighters were GPS-equipped.[38] Commercial receivers were the only option for coalition aircraft, so British pilots attached handheld GPS units with Velcro to the instrument panels of their Jaguar fighter-bombers.[39] Pilots also had to ensure that their GPS receivers and inertial navigation systems used the same map coordinate standards, or bombing errors resulted.[40] Whether the GPS was permanently installed or improvised, GPS-equipped planes constituted a tiny fraction of some 1,600 attack aircraft used by U.S. and coalition forces, but they led others in their formations to precise launch coordinates, which in some cases reportedly doubled bombing accuracy.[41]

GPS capability influenced planning for the first strikes of the air campaign—raids to destroy two Iraqi early warning radar sites near the western border, opening a window for squadrons of strike fighters headed deep into Iraq. Planners lacked detailed maps of the target locations, and they wanted eyewitness confirmations that the attacks had achieved complete destruction—something long-distance missile strikes alone would not provide.[42] Four Air Force Special Operations Pave Low helicopters equipped with GPS led eight Army Apache attack helicopters on terrain-hugging, under-the-radar flights across the desert. Since the tightly drawn mission plan required hitting two sites forty miles apart within twenty seconds of each other, the task force leader used the time from a GPS satellite's atomic clock to call out the synchronization mark for all cockpit clocks as the mission began.[43] The Pave Lows dropped chemical glow sticks on the ground about ten miles from each site, marking precise points where they broke off their lead and the Apaches inputted predetermined targeting coordinates for their Hellfire missiles.[44] The Apache crews followed up the Hellfire strikes by hitting the sites with one hundred smaller, unguided rockets and four thousand rounds of cannon fire to finish the job before darting back toward the border as coalition jets screamed overhead toward Baghdad.[45]

GPS improved the military's ability to locate and retrieve troops and equipment caught behind enemy lines. At the start of the ground campaign, an eight-man Army Special Forces team was inserted deep into Iraq, about two hundred miles south of Baghdad, to monitor military traffic on a highway. Discovered by civilians and later engaged by more than 150 Iraqi soldiers, the team requested air cover and an "emergency exfiltration," which arrived in time to save everyone thanks to having precise coordinates of their location.[46] In a similar incident that same day a Navy Seal team was being picked up after completing a mission off the coast of Kuwait when they accidentally dropped an expensive piece of equipment into the water. Stopping to retrieve it risked detection and compromising the mission, so they recorded the precise location using GPS coordinates and returned the following night to recover it.[47]

Navy vessels used GPS not only to navigate to specific launch points as required for the TLAM's terrain-matching system but also to accurately map and clear extensive minefields. Saddam Hussein, believing that air power was not decisive in wars and assuming that coalition troops would find maneuvering in the desert as difficult as his own forces did, expected an amphibious invasion. A large full-color relief map discovered after the war in a Kuwaiti building the Iraqis converted to a military post confirmed that Hussein had committed seventy thousand to eighty thousand troops and up to half of his artillery to defend the Kuwaiti coast.[48] Iraq turned oceanfront high-rise condominiums into gun perches, littered the beaches with earthen berms, trenches, mines, barbed wire, and other obstacles, and placed more than 1,200 sea mines in a 150-mile arc from the Iraqi-Kuwaiti border southward to the Saudi Arabian border.[49] Navy and coalition ships conducted minesweeping operations prior to the ground offensive as part of a "right feint" maneuver designed to convince Hussein he was right and spent a year after the cease-fire destroying mines and removing war wreckage to reopen commercial shipping lanes.

Space Facilitates Ground Offensive

The right feint's counterpart, the western "left hook" flank attack, which Gen. H. Norman Schwarzkopf, commander in chief of the U.S. Central Command, compared to a Hail Mary play in football, vividly illustrates the impact of GPS during Desert Storm. Soon after knocking out Iraq's electronic communications and grounding its air force planes or sending them fleeing to Iran, Schwarzkopf ordered more than two hundred thousand troops and tens of thousands of vehicles massed on the Saudi-Kuwaiti border to move, undetected, more than two hundred miles to the west. The westward shift positioned two full U.S. Army Corps, the XVIII Airborne Corps and the VII Corps, pulled from Cold War bases in Germany, to bypass and encircle the bulk of Iraq's dug-in positions. This enormous march of men and materiel spanning three weeks did not itself require GPS navigation. The entire convoy rolled northwest on a narrow but straight two-lane blacktop road adjacent to the Trans-Arabian Pipeline. Once the troops, armored equipment, and logistical supply units left the roadway and prepared for desert battle, however, GPS navigation, together with satellite communications and imagery and night-vision devices, became indispensable. Never before had armies waged battle on such a large scale, both day and night in hostile desert conditions.[50] GPS helped advancing units maintain alignment with others on each flank, helped avoid fratricide through mistaken engagement with friendly forces, aided combat search and rescue, and helped ensure the delivery of food, fuel, and supplies where needed on a timely basis.[51]

This new situational awareness of the location of friendly troops marked the birth of what became Blue Force Tracking (blue for friendly forces, red for enemies). After the war, work began on systems like the Force XXI Battle Command Brigade and Below (FBCB2) program. Combining GPS, a satellite phone, a laptop computer, and special software, the technology displays maps with blue icons pinpointing friendly units and updates their positions automatically in real time. Users can communicate by radio, text message, or e-mail. Blue Force Tracking expanded rapidly in 2003 during Operation Iraqi Freedom and has evolved into subsequent generations known as Joint Capabilities Release and Capabilities Set 13, with more than one hundred thousand devices installed in U.S. armored vehicles, tanks, and helicopters.[52] Although Schwarzkopf does not use the term *Global Positioning System* in his autobiography, *It Doesn't Take a Hero*, published a year after the war, he mentions the technology in a reference to the challenges of desert warfare. He writes, "In Europe, soldiers had been able to orient themselves in relation to roads, towns, forests, and other landmarks; in the desert, there were no landmarks and even the dunes moved. So we had to quickly teach the use of *satellite navigation equipment*, celestial navigation and dead reckoning" (emphasis added).[53]

Celestial navigation, of course, requires clear weather, but the Persian Gulf War occurred during the worst conditions recorded in the fourteen years the Air Force had been keeping records of Iraqi weather.[54] Bad weather forced cancellation of 15 percent of planned air sorties during the first ten days of the air campaign and affected operations throughout the war.[55] The ground campaign began February 24 amid rain and sandstorms and after a tug-of-war between the generals in the field, who wanted to wait for better weather, and civilian leaders in Washington, who were managing last-minute negotiations with the Soviet Union that could have resulted in Saddam Hussein giving up Kuwait but keeping his massive military apparatus. Absent GPS, the outcome of the war may not have been different, but its duration would certainly have been longer.

The authors of a popular Schwarzkopf biography, *In the Eye of the Storm: The Life of General H. Norman Schwarzkopf*, published in late summer 1991, describe vividly the battlefield advantages of GPS. Their use of the plural *systems* also highlights the still-evolving use of the term at the time, as well as the trend toward conflating the receivers with the entire system:

This element of surprise was compounded on the second day of the war by an appalling sandstorm that reduced visibility to less than a hundred feet.

The storm created problems for the coalition. For example, some aerial resupply missions had to be stopped. But it created far more problems for the disoriented Iraqis. Once again, superior technology aided Schwarzkopf's forces. Using the navigation devices called global positioning systems, they always knew exactly where they were and where they were heading. These high-tech gadgets, based on digital coordinates of positions and satellite connections, could tell soldiers how far they were from pre-set targets even as sand blew sideways into their faces. Rampaging out of whipping sand on Iraqis looking in the other direction, or, later, out of the sight of soldiers who did not even know the coalition was inside Iraq, the allied forces had an effect that was indeed as sudden, dramatic, and enveloping as thunder and lightning.[56]

Improvisation Saves the Day

Before troops could employ GPS in the conflict, the military had to overcome several problems. Foremost was the shortage of receivers and accurate maps. Because the system would not be complete until 1993, the services had not yet acquired many receivers, much less trained personnel widely in their use. The Army owned just five hundred demonstration receivers when troops began deploying to Saudi Arabia for Operation Desert Shield.[57] Despite general unfamiliarity with GPS, its impact was immediate and troops quickly embraced the new technology, as illustrated by the response of an assistant signal officer from the 11th Air Defense Brigade when asked about GPS receivers: "If you mean those green position locators, they are lifesavers. Whenever we sent someone to another unit for coordination, we entered that unit's ten-digit coordinates and the SLGR (small, lightweight GPS receiver) directs them to the command post. Before, we had people getting lost in the desert, but since we got the three GPS receivers, nobody has got lost."[58]

The SLGR, or "Slugger," a version of Trimble's Trimpack receiver discussed in the preceding chapter, was the most widely fielded receiver in the Persian Gulf War, with about four thousand ultimately deployed.[59] Although it lacked the ability to decode the more precise, encrypted (P code) signals used by the military's AN/PSN-8 "Manpack" GPS receiver built by Rockwell Collins, it was more versatile. The Manpack, about the size of two stacked shoeboxes, weighed seventeen pounds and was portable in the sense that a soldier could wear one on a special backpack-style frame. In practice, it more often was lashed to a wheeled vehicle or helicopter.[60] The SLGR weighed three and a third pounds and was about two inches thick by seven inches square—small enough and light enough to hold in one hand. With the switch of a button it could display coordinates in several different mapping grid systems, making it usable for aviation, ground movements, and artillery support.[61] The difference in cost between the SLGR and Manpack exceeded their size and weight differences. A SLGR cost about $3,400, whereas the Manpacks cost $45,000 apiece.[62] The Magellan NAV 1000, smaller still and battery-powered, was the second most popular receiver, with about one thousand used in the Persian Gulf War.[63] Commercial receivers provided accuracy to within about eighty-two feet in each direction (latitude, longitude, and altitude) compared to the fifty-three-foot accuracy of military receivers.[64] The military later developed a receiver known as the PLGR (precision lightweight GPS receiver), which decoded the encrypted signal and was resistant to jamming (radio interference that blocks the GPS signals) or spoofing (false signals broadcast to cause receivers to generate incorrect coordinates).

Soon after the first troops arrived in Saudi Arabia, the Pentagon began placing orders for commercial receivers and eventually ordered about ten thousand. Some service members asked their families to ship them commercial units. The demand strained production capacity, and by the war's end, the number deployed by all branches of the military totaled 4,490 commercial and 842 military receivers, with the Army absorbing about four-fifths of the commercial versions and two-thirds of the military ones.[65] Coalition partners began the conflict with about two dozen receivers and eventually fielded more than two thousand of the devices, which Arab ground commanders dubbed "the magic compass."[66] As with aircraft, U.S. ground commanders apportioned available receivers where they thought they could do the most good. This meant, for example, divvying up three thousand receivers among forty thousand vehicles in the VII Corps alone. Receivers typically went to reconnaissance teams, artillery surveyors, and the forward units leading the charge.[67]

Recognizing immediately that GPS would be vital and most receivers could not use the encrypted military signal, the Pentagon on August 10, 1990—eight days after Iraq invaded Kuwait—disabled Selective Availability, the intentional degrading of the civilian signal to make it less accurate. The U.S. Air Force Space Command, which manages the satellites and ground control stations, had first activated the security feature only five months earlier, on March 25. Commanders reasoned that Iraq lacked weapons that could use GPS. However,

Fig. 7.3. Manpack GPS receiver by Rockwell Collins. (Courtesy Rockwell Collins)

Fig. 7.4. AN/PSN-11 PLGR by Rockwell Collins. (Courtesy Rockwell Collins)

the move opened the possibility of Iraqi troops acquiring and using commercial receivers for their own desert navigation, a fear that never materialized. The Air Force reactivated Selective Availability in July 1991, but within a decade turned it off again after a policy debate (see chapter 8).

Anyone who has failed to update the software in their automobile navigation system and turned onto a new highway not shown on the screen has experienced the disorientation that using GPS with outdated maps can produce. When soldiers easily found other units in the empty desert during the Gulf War, they knew their destination's coordinates. Without accurate, known coordinates, GPS could not lead troops to enemy targets. For this need non-GPS satellites, both military and commercial, made a critical difference in the war. Most Middle East maps the Defense Mapping Agency had available in August 1990 dated from the 1960s up to 1983.[68] They used map reference systems predating the World Geodetic System 1984 (WGS-84), which GPS uses. (Cartographers periodically revise the underlying frame of reference, or datum, they use in mapmaking as methods for precisely measuring the planet improve.)

B-52 bombers using GPS saw their gravity bombs fall consistently four hundred to six hundred feet short of their targets before figuring out that their inertial navigation systems were set to WGS-72, the older system from 1972.[69]

To solve the map problem, the Defense Department contracted with two private companies, American-owned Earth Observation Satellite (EOSAT) and SPOT Image, a French firm, to downlink wide-area digital satellite images to receiving stations in Saudi Arabia. This arrangement enabled military map-makers to create updated maps of the swath of land under a satellite pass within a matter of hours and distribute paper copies to troops in the field. These satellite images were not simply high-powered photographs made from space. They were multispectral images, derived from sensors capable of measuring wavelengths not visible to the eye. Multispectral imaging can detect heat and moisture, revealing water depth along coastlines, heated structures beneath vegetation, and tracks where equipment has traveled over the ground.[70] EOSAT's two Landsat satellites carried onboard GPS receivers, yielding precisely known orbits and thus images with highly accurate coordinates ready-made for use with GPS receivers.[71] In the hunt for Iraqi SCUD mobile missile launchers, reconnaissance teams combined these assets with sensor data from three U.S. spy satellites of the Defense Support Program (DSP). The Pentagon began developing DSP satellites in the late 1960s to detect ballistic missile launches, primarily Soviet ICBMs. Desert Storm constituted the first "real-world" test of the system, and although it was developed as a strategic asset, military analysts praised its value for tactical battlefield use.[72] This use of satellite technology and digital image processing to provide real-time space surveillance of enemy movements was a technological milestone of the war. Together with GPS navigation, the overwhelming use of satellites led military leaders to proclaim the Persian Gulf War "the first space war."[73]

In a notable historical coincidence, Saddam Hussein launched his invasion of Kuwait on the same day, August 2, 1990, that the Air Force launched a GPS satellite into orbit. The Air Force launched two more satellites over the next three months, one on October 2 and another on November 26, bringing the constellation size to sixteen—still five short of the minimum required for continuous, worldwide three-dimensional coverage. Although some accounts erroneously describe these satellites as hurried into orbit to prepare for battle, the launches went up according to a schedule established well before the invasion.[74] Neither the Air Force nor its suppliers could accelerate the launch program, given its complexity.[75] However, the military took several steps to optimize the incomplete constellation for use in Iraq. Before the October launch, Space Command altered that satellite's orbit to increase its visibility over the Persian Gulf and shortened the testing and preparation phase to get it into service faster.[76] Tinkering with the constellation placed three satellites in view for nearly round-the-clock two-dimensional coverage (latitude and longitude, needed for ground and ocean-surface operations) and provided four satellites in view for three-dimensional coverage (latitude, longitude, and altitude, needed for airplanes) for about eighteen hours a day.[77] At the time, three-dimensional coverage over Fort Bragg, North Carolina, for example, was available only fifteen hours per day.

During the crisis three satellites experienced problems that kept technicians of the 2nd Satellite Control Squadron (now called the 2nd Space Operations Squadron) busy designing workarounds.[78] In August 1990 a twelve-year-old satellite (the third experimental Block I satellite, launched in October 1978) lost 40 percent of the power from its solar panels. By shutting down GPS signal transmissions and nonessential power uses during periods when the solar panels were in the earth's shadow, ground controllers kept the batteries recharged and providing coverage over the Persian Gulf.[79] On December 10 a stabilizer on another older satellite (the sixth Block I satellite, launched in April 1980) malfunctioned, and technicians devised a way to keep its solar panels angled toward the sun and its antennas pointed toward the earth.[80] Ground crews kept it partially working during the war, but they permanently removed it from the constellation in March 1991.[81] Also in December, the solar panels aboard the newest satellite, just launched November 26, became stuck in a fixed position, and ground controllers had to adjust them manually for the remainder of the war.[82]

In sum, the Persian Gulf War served as a sort of beta test for GPS-guided munitions but offered more of a full dress rehearsal for the system as a navigational aid for foot soldiers, ground vehicles, ships, and aircraft. For stateside ground crews managing the sometimes-balky satellites, it was a time of outright experimentation similar to what NASA faced during the moon race.

After the war the Pentagon concluded, "GPS was used more extensively than planned and met navigation and positioning requirements. . . . GPS should be considered for incorporation into all weapon systems and platforms."[83] Those military leaders who argued in 1981 that "it may be difficult to understand the full potential until the system is deployed and the vast number of potential users are able to see what it will do for them" undoubtedly felt vindicated.[84] The war's short duration and minimal casualties surprised almost everyone and left a legacy that makes it difficult to recall today the dire predictions and street protests that preceded it. A protester in front of the White House on the day the air campaign started questioned whether the carnage would leave anyone to liberate, remarking, "Bombs have no eyes."[85] But many did have eyes, and many more bombs with eyes would follow. The Persian Gulf War marked a turning point in public expectations about acceptable collateral damage in conflict. The services rapidly began integrating GPS capability into precision-guided weapons, which saw increased use in each subsequent conflict. Whereas precision weapons constituted a negligible percentage of all ordnance expended in Vietnam, they represented 7 percent in Desert Storm; the share expanded to 60 percent during Operation Deliberate Force in the Balkans in 1995 and grew to 80 percent during Operation Allied Force in Yugoslavia and Kosovo in 1999.[86]

GPS applications continue to spread to "all weapons systems and platforms," including field artillery. In early 2012 General Dynamics and BAE Systems announced the successful demonstration of a "smart" 81-mm precision mortar round made possible with GPS guidance and an in-flight control system.[87] What could be next—a smart bullet?

The actual size of that "vast number of potential users" first became evident in 1991. Others beyond the military took note of how GPS succeeded in the Persian Gulf War. The lopsided ratio of commercial to military receivers in Desert Storm foreshadowed a numerical difference that would become permanent.

Going Mainstream
A Consumer Industry Is Born

8

Time is the greatest innovator.

Francis Bacon, "Of Innovations," The Essays, *1625*

Hearing no more gunshots, voices, or footsteps for about an hour, Capt. Scott O'Grady decided to risk calling for help on his survival radio.[1] He had spent most of the past three hours lying motionless, facedown in the dirt, with his gloved hands covering the exposed sides of his face and ears. The events that landed him in the Bosnian underbrush, hiding like a hunted rabbit, replayed in his mind, interspersed with reminiscences from his youth, lessons from his Air Force training, and scenarios for his survival. One moment he was cruising at twenty-six thousand feet, the next he was riding a fireball as his cockpit disintegrated around him. Four seconds after his F-16 fighter's alarm system warned that targeting radar had locked onto him, a Serbian SA-6 surface-to-air missile slammed the plane's belly, slicing the aircraft in two. He recalled spotting the yellow ejector-seat handle and tugging it, thrusting himself into the clouds and away from the burning debris that was traveling 350 mph. His long descent by parachute, observed by Serbian soldiers, left him only minutes to free himself from its harness and run from the grassy clearing where he landed to the cover of nearby woods. As their prey, he must be cautious with his survival radio lest its signal lead them to him before rescuers could arrive. He pressed the button and whispered his call sign, "Basher Five-Two," but all he heard was static.

O'Grady steeled himself for the possibility that he might have to spend days or weeks evading capture. He was the second pilot shot down enforcing the United Nations no-fly zone over the Balkan civil war, but he could not rely on the locals returning him to friendly forces as they had done previously for a British Harrier jet pilot. As the first U.S. pilot shot down since Desert Storm, he knew members of his 555th Fighter Squadron out of Aviano, Italy, the entire U.S. military, and much of NATO would be combing the skies, the airwaves, and intelligence sources for any sign he had survived. He needed to figure out his location and make his way to higher ground, where his radio would have a better chance of reaching search-and-rescue aircraft and where helicopters could more easily land. He must make that journey only in darkness, hiding under camouflage by day. O'Grady reached into the pocket of his survival vest and pulled out his GPS receiver, a Flightmate Pro made by Trimble Navigation.[2] The three-by-seven-inch handheld unit operated on four AA batteries, but its four-line, backlit liquid crystal display was bright enough that he cupped his hands around it to keep the glow from piercing the darkness. Once the receiver acquired signals from three satellites, he had his position and oriented himself on a laminated topographic map from his survival kit. His target, a large hill, was about two miles south, but his stealthy advance, confined to the hours between midnight and 4:00 a.m., consumed that night, June 2, 1995, and the next four nights. He subsisted on rainwater, grass, leaves, and insects. At 2:07 a.m., June 6, he finally made radio contact with a fellow F-16 pilot from Aviano who was straining the jet's fuel supply crisscrossing the skies in hopes of hearing something. Using his GPS unit, O'Grady was able to provide accurate coordinates. Military leaders rapidly mounted a rescue operation. Four and a half hours later, a platoon of Marines specially trained in tactical rescues used GPS guidance and extraordinary piloting to land two seventy-foot-long CH-53 Super Stallion helicopters in a small, fog-shrouded clearing about two hundred yards from O'Grady. He ran out of the woods, and they hauled him aboard one of them. The Super Stallions flew fast and low on the way out, drawing antiaircraft and small-arms fire. At least two shoulder-fired missiles whizzed behind them. Just before 7:30 a.m., the choppers landed on the USS *Kearsarge* in the Adriatic Sea to a hero's welcome—one that would be repeated later at Aviano and again at the White House.

O'Grady's dramatic story of survival and rescue offered the media and viewing public a welcome break from the O. J. Simpson murder trial and a grinding Medicare debate in Congress. More than half of people surveyed at the time said they were still closely following news about the Oklahoma City bombing two months earlier.[3] Against this gloomy backdrop, O'Grady became an instant sensation. "High-Flying O'Grady Fills Hunger for Hero," read a USA Today headline above a story about how the crush of requests from reporters, agents, authors, and filmmakers was keeping all sixty Air Force public affairs staffers busy.[4] Media interest endured, with additional bursts of attention when O'Grady later released a book and again on the one-year anniversary. News executives voted O'Grady among the top ten news stories of 1995.[5] The coverage meant lots of publicity for GPS, but as reporters dug for different angles some asked why it took six days to find the downed pilot. After a private meeting with O'Grady, Rep. Robert Dornan, a California Republican who was chairman of the House Intelligence Subcommittee on Technical and Tactical Intelligence, declared O'Grady's PRC-112 survival radio obsolete.[6] Dornan, an ex–fighter pilot himself, revealed that O'Grady told him if he ever had to bail out again he would like to have a couple of cellular telephones.[7] British pilots reportedly were already carrying them.[8]

The O'Grady search-and-rescue effort, reminiscent of incidents in Vietnam thirty years before, prompted calls for immediately funding a stalled Air Force program to replace the outdated PRC-112 with a digital radio capable of satellite communications.[9] By fall 1995, Motorola was working under a "quick-reaction program" to add GPS to the PRC-112, creating a model known as the Hook 112, but those units still relied solely on less powerful "line-of-sight" communications.[10] In February 1996 the GPS joint program office awarded Rockwell International a $13 million contract to build eleven thousand Combat Survivor Evader Locator (CSEL) radios for delivery by 1998.[11] Contract options provided for twenty-seven thousand radios by 2001 at a cost of $67.2 million.[12] CSEL development took longer than expected, and in 1999 Department of Defense officials deemed the first radios "not operationally suitable."[13] Refinements continued. In 2004 the military began fielding the CSEL radios, first proposed in 1991, at a cost of about $5,000 apiece.[14] By 2011, Boeing (which acquired Rockwell International's aerospace and defense businesses not long after the initial CSEL contract was awarded) delivered fifty thousand radios for use by the Army, Navy, and Air Force.[15] Wars in Afghanistan and Iraq with hundreds of thousands of troops and many civilian contractors on the ground enlarged the radio's mission well beyond locating downed pilots to include general "personnel recovery."[16] CSEL radios, sometimes called the military's "global 911," combine a precision-code GPS system, a locator beacon, over-the-horizon satellite data communications, and line-of-sight voice communications.[17] The GPS coordinates and secure digital signal allow search-and-rescue teams to locate and authenticate missing personnel even if they cannot make voice contact, prompting CSEL program managers to coin the slogan "no search and all rescue."[18]

The military's search-and-rescue needs, illustrated by the O'Grady episode, provided the impetus to combine compact mobile communications with precision GPS tracking of receivers. Public safety concerns and commercial motives later drove a similar evolution in civilian telephony. In the same way that development of increasingly accurate atomic clocks had made precise satellite navigation feasible decades before, technological advances paved the way for GPS to become a ubiquitous public utility. One was the evolution of techniques to augment the accuracy of GPS signals, providing more precise and diverse applications; another was miniaturization of GPS receivers onto a single computer chip, enabling their migration to and integration with other devices. However, technological progress by itself was not enough. The private sector faced uncertainty.

Growing Public Policy Issues

In early June 1995, while Scott O'Grady was evading capture and using GPS to aid his rescue, Congress and the Department of Defense had just received a report from a yearlong study concerning the future of the technology, jointly prepared by the National Association of Public Administration (NAPA) and the National Research Council (NRC).[19] Simultaneously, the White House Office of Science and Technology Policy and the National Science and Technology Council were awaiting delivery of a similar study undertaken by RAND.[20] Leaders commissioned these studies to plot the direction of GPS policy, recognizing it had not kept pace with technology.

As commercial interest in GPS surged following the Persian Gulf War, policy makers realized that the system's dual-use nature posed fundamental challenges for balancing national security needs with the enormous economic opportunities it offered. Ongoing U.S. military control of GPS was a source of tension for civilian users, both domestic and abroad. Some thought the government should privatize it partially or completely. Others envisioned oversight by a multinational civilian agency akin to the International Civil Aviation Organization (ICAO).[21] Solving these issues was difficult because they crossed not only borders but also traditional bureaucratic boundaries, and there was no mechanism for sharing information and responsibilities across the different missions of the Departments of Defense, Commerce, and Transportation, a problem often described as "stove-piping."[22]

One lesson the Gulf War taught about satellite systems was that "everyone will want them," Martin C. Faga, an assistant secretary of the Air Force, told a U.S. Space Foundation symposium in April 1991.[23] Already the Soviet Union had launched much of its similar GLONASS constellation (now a Russian system, following the Soviet breakup in December 1991). The European Union (EU), dependent on two defense-run systems, began to contemplate building its own system. Before the end of the decade, in May 1999, the EU authorized the first funds for its commercially oriented Galileo system, described initially as a twenty-one-satellite constellation similar to GPS.[24] Therefore, concerns in the early 1990s about how the United States should manage cooperation and competition with other global navigation satellite systems were quite valid.

From an economic standpoint, many feared that the policy of degrading the civilian signal and the military's ability to shut down GPS service at its discretion would suppress private investments in research, development, and manufacture of commercial products or consumers' willingness to buy and integrate those products into their business or personal activities. Domestic commercial GPS sales were $213 million in 1991, and the U.S. GPS Industry Council projected a 62 percent rise by 1996.[25] Farmers had already begun experimenting with GPS to spread expensive fertilizers more accurately, boosting yields while reducing costs.[26] The Department of Transportation estimated that traffic congestion caused $73 billion in lost productivity each year, highlighting the early opportunities that many companies saw in the automobile navigation market.[27] From tracking endangered species to fleets of rental cars and delivery trucks, GPS empowered a surge in geographic information systems (GIS), a $1.41 billion worldwide market in 1990 that the market research firm Dataquest projected would double by 1994.[28] Some companies with large communications infrastructures were already adopting GPS timing to synchronize their data networks, but Atlanta-based Southern Company, a power and electric utility operator, specifically cited the military's ability to degrade the signal when it chose in 1991 not to use GPS to time its networks in five states.[29] Private investors also had to contemplate whether the government would reliably maintain the system over the long term and whether service would remain free of charge or be subject to taxes or user fees.

From the military's perspective, national security concerns trumped potential domestic or worldwide economic benefits. Following the successful debut of GPS in the Persian Gulf War, U.S. military leaders feared that enemies would soon take advantage of the signals for hostile use. GPS capability offered Third World adversaries an inexpensive way to upgrade SCUD missiles or even to launch what the Pentagon termed "the poor man's cruise missile" against the United States or U.S. interests abroad.[30] The strategy to counter this threat was to carefully limit access to the more accurate, encrypted, and jam-resistant military signal (the Precise Positioning Service, or PPS) and to "dither" the civilian signal (the Standard Positioning Service, or SPS), rendering it less accurate. In late August 1991, a few months after the Air Force reactivated Selective Availability following the war, the U.S. government revised its international trade regulations to reflect the dichotomy, classifying civilian receivers as unrestricted "general destination items" but maintaining tight export restrictions on military-grade receivers, treating them as "munitions."[31] By December 1993, the Department of Defense reported that about eleven thousand military-grade GPS devices had been sold or provided to allies, with nearly nine thousand going to Europe and hundreds each to Japan, Israel, Canada, South Korea, and Australia.[32]

Even as the military degraded the civilian signal, Pentagon officials acknowledged "an inherent quandary" posed by simultaneous multimillion-dollar programs the Coast Guard and Federal Aviation Administration pursued to boost SPS accuracy for enhanced safety around seaports and airports.[33] These local augmentation systems used the technique known as differential GPS (DGPS), which employs GPS receivers and ground-based radio beacons at fixed locations with precisely surveyed latitude, longitude, and elevation coordinates. The reference stations make it possible for computers to detect tiny inaccuracies in the signals from space, calculate the differences, and transmit corrections to users via the ground beacons. At that time Selective Availability reduced the civilian signal's accuracy to within about 328 feet (100 meters), and the precise military signal was accurate to within about 38 feet, but DGPS could provide accuracies within about 10 feet.[34] While the military could presumably shut down DGPS systems serving U.S. territory in the event of an attack, it resisted plans backed by the FAA and the International Air Transport Association to build a Wide Area Augmentation System (WAAS) that would use multiple reference stations and separate non-GPS communications satellites to broadcast corrected signals to airplanes across the entire continent.[35]

As the Pentagon and Department of Transportation tried to resolve their differences, U.S. officials repeatedly sought to reassure the private sector, particularly the huge civil aviation market. The FAA administrator, in a letter to the ICAO's Tenth Air Navigation Conference in September 1991, pledged to make the Standard Positioning Service available to the international community at no charge for at least ten years (starting in 1993).[36] Other than specifying a time frame, this was essentially the same offer President Reagan had made a decade before. The FAA repeated that pledge at ICAO assemblies in 1992 and 1994, assuring provision of SPS for the "foreseeable future," subject to available funds, and additionally promising to provide at least six years of advance notice before terminating GPS or eliminating SPS.[37]

All of this activity occurred before GPS was officially complete. Not until June 1993, almost twenty years after the Defense Systems Acquisition Review Council authorized the program, was the full constellation of twenty-four satellites orbiting the earth. Following several months of testing, Secretary of Defense Les Aspin declared Initial Operational Capability (IOC) in a December 8 letter to the Department of Transportation. After nearly a year and a half of further testing, the Air Force declared Full Operational Capability (FOC) in April 1995.[38] These actions were not mere ribbon-cutting-style announcements; they were formal certifications that the system met specific criteria for accuracy and reliability—for example, accurate horizontal positioning within one hundred meters 95 percent of the time. The FOC declaration assured authorized users of the Precise Positioning Service that PPS met its more-stringent requirements, which was of significance to U.S. military forces and allies. The IOC certified that SPS met standards set forth in the Federal Radionavigation Plan, a technical planning and policy document prepared jointly by the Defense and Transportation Departments every two years. One practical effect was to designate GPS receivers as satisfying federal maritime regulations requiring ship-borne electronic position fixing devices.[39] Such devices enable vessels to report their position manually or automatically to avoid accidents, such as in fog. In March 1999 the Coast Guard activated an augmentation system called the Maritime Differential GPS Service. It consisted of a control center and network of radio beacons for enhanced navigational accuracy along the coastlines of the continental United States, the Great Lakes, Puerto Rico, the U.S. Virgin Islands, portions of Alaska and Hawaii, and much of the Mississippi River basin.[40]

Approving GPS for Aviation

The FAA began phasing in GPS use in June 1993, six months before the IOC declaration.[41] Pilots could immediately use approved GPS equipment as their primary en route navigation method for domestic and oceanic routes, but its use with landings came with restrictions. At first pilots could use GPS only for nonprecision approaches—instrument landings that use ground-based radio signals to align the plane with the runway but do not provide vertical guidance, which remains visual. By contrast, landings with reduced visibility require precision approaches, where ground equipment transmits both lateral and vertical guidance. Airports use a variety of navigation aids, and some lack instrument landing systems that can provide vertical guidance. The initial FAA phase-in opened five thousand GPS approaches at 2,500 U.S. airports.[42]

Formal FAA approval of GPS for civil aviation followed on February 17, 1994, and the next month the agency published the first certified GPS approaches—a process that would take years to cover all affected airports.[43] Published approaches originally filled thick paper manuals, but with the advent of digital circuitry they were programmed into cockpit instruments, including GPS receivers. Until the FAA certified a particular runaway approach for GPS, pilots had to follow the traditional approach and actively monitor existing ground-based navigation aids unless the plane was equipped with GPS instruments featuring Receiver Autonomous Integrity Monitoring (RAIM) or an equivalent method of detecting faulty satellite signals.[44] RAIM circuitry and software enables receivers to verify the accuracy of a position derived from four satellites by comparing it to signals from a fifth or sixth additional satellite.[45] Despite the requirement for backup systems, FAA officials left no doubt that GPS was the direction for the future. At the time, FAA administrator David Hinson remarked that he saw no reason GPS would not become the only navigation system.[46]

Hinson's comment reflected a widely held view: the arrival of GPS represented enormous potential cost savings for the general aviation sector, airlines, airports, and the overburdened FAA-managed air traffic control system. Some seven hundred thousand general aviation pilots operating two hundred thousand aircraft looked forward to the greater affordability of panel-mounted GPS receivers costing $2,000 to $5,000 that would eventually replace $20,000 worth of traditional navigation equipment.[47] Continental Airlines in 1993 was the first carrier to seek FAA approval for GPS-only approaches into Aspen and Steamboat Springs, Colorado, where weather conditions often forced cancelation of a quarter of the airline's scheduled flights or diversions to alternate airports.[48] The airline calculated that its investment in Honeywell flight management systems with RAIM-enabled GPS equipment for its regional jets would achieve payback in less than one year.[49] For Honeywell, it was the first sale of the system to a regional carrier—a large potential market; for the FAA, the early success fueled hopes that the new technology could help contain the rising costs of extensive ground-based instrument landing systems and even extend precision approaches to airports that lacked them.[50] The total number of airports and landing areas of all types in the nation grew by 15 percent from 1980 to 1990, when there were 17,490 facilities.[51] By 1994, there were 853 more.[52] From the time Congress deregulated the airline industry in 1978, the number of passengers boarding aircraft rose 270 percent to about 465 million in 1990.[53] That figure surpassed 528 million by 1994, driven by fare wars that erupted in 1992.[54] As major carriers abandoned smaller cities in favor of the hub-and-spoke system, smaller regional carriers moved in. These commuter airlines strained the FAA's inspection regime, accidents among them spiked, and the GAO issued a report that prompted the FAA to hire more inspectors and refocus its attention on regional carriers.[55] Despite increased volume, airline travel became a commodity with razor-thin or zero profits, forcing carriers to cut costs and achieve greater efficiencies wherever possible.[56]

A Continental executive, testifying before the House Subcommittee on Technology, Environment, and Aviation in March 1994, estimated that GPS navigation would allow his company to capture $1.9 million annually in revenue otherwise lost due to weather-related flight cancellations.[57] Beyond fewer cancellations, he said, GPS navigation could save the airline industry $5 billion annually through reduced delays and more direct routing.[58] With that figure in mind, and citing projected industry-wide spending of $500 million to $740 million on new GPS-enabled avionics, he urged the government to speed development of the FAA's proposed Wide Area Augmentation System.[59] FAA associate administrator Martin Pozesky told the same subcommittee that his agency planned within months to authorize privately owned local-area DGPS systems at airports to provide precision approaches under Category I weather conditions.[60] Category I instrument landings require a pilot to be able to see the runway well enough to decide whether to land at a range of at least 1,800 feet and a height no lower than 200 feet above the runway.[61] Category II and III approaches, denoting poor visibility, require much more precise instrument guidance, as they place the pilot closer and lower before he spots the runway, leaving little margin for decision making. Pozesky told the subcommittee that WAAS could provide Category I landing guidance at virtually all major airports across the United States without the need to install local ground-based DGPS systems.[62] At the time officials estimated that about one hundred existing approaches would need differential ground systems for more precise Category II and III landings.[63] The decision to back WAAS also signaled a move away from microwave landing systems, a competing technology that many European nations continued to pursue.[64]

By mid-1994, the FAA began soliciting bids for WAAS, anticipating that it could begin fielding the system in mid- to late 1997 and complete it within six years for between $400 and $500 million.[65] The design called for about two dozen ground reference stations across the nation and a central station for sending the signals via satellites to airplanes. Five teams, each a consortium of companies, bid on the project, but when the time came to award the contract early in 1995, the announcement was delayed; the FAA had learned the Pentagon was planning new experiments to jam the same signal it was trying to enhance. After months of wrangling, the Transportation and Defense Departments announced they would conduct "joint testing and review of WAAS," but with the civil aviation industry unnerved by the episode, this time President Bill Clinton himself sent a letter to the ICAO reaffirming the U.S. commitment to civil use of GPS.[66]

Crafting a Policy Framework

It is not surprising that the study findings delivered two months later by the National Association of Public Administration and the National Research Council included in capital letters the plea, "THE UNITED STATES NEEDS A NATIONAL STRATEGY FOR GPS."[67] The NAPA/NRC study recommended that the president adopt specific national goals and create an executive board to implement them. Similarly, the RAND study suggested a presidential decision directive, a type of executive order, as the policy-making framework. The two studies' conclusions and recommendations had more in common than they differed. Both stipulated the priority of national security but saw great risks in failing to engage international users and manufacturers. Of the top eleven companies holding international patent rights for GPS products in 1994, four were Japanese and one was French. All were well-known names: Pioneer Electronic, Motorola, Mitsubishi Denki KK, Caterpillar, Trimble Navigation, Hughes Aircraft, ITT, Magnavox, Nissan Motor KK, Sony, and Thomson CSF.[68] Both studies emphasized maintaining U.S. leadership in satellite navigation systems and advancing GPS as the internationally accepted standard. That required minimizing barriers to trade, broadening civilian participation in policy making, and continuing to provide GPS free of direct user fees. However, neither study saw an alternative to keeping GPS—the space and ground segments—under military control, both for security reasons and because of the need for stable funding. The user segment—receivers, applications, and services—was already "effectively in the hands of the private sector," the RAND study noted.[69]

Both studies offered eye-catching market projections for GPS equipment sales. The RAND study cited a U.S. GPS Industry Council estimate of 50 percent annual growth that would carry worldwide sales from $510 million in 1993 to $8.47 billion in 2000.[70] Car navigation devices and GPS-equipped computers and mobile phones accounted for more than 60 percent of that total, while military sales represented just 1 percent. The NAPA/NRC study surveyed seventy-nine companies and arrived at an even more optimistic forecast, projecting that an estimated $2 billion global market in 1995 would grow to at least $11 billion by 2000 and $31 billion by 2005.[71] One chart in the NAPA/NRC study suggested that turning off Selective Availability could significantly boost commercial growth.

The two sets of recommendations differed most over the contentious policy of degrading the civilian signal. The NAPA/NRC study recommended turning off Selective Availability immediately and deactivating it after three years. It based this recommendation on economic projections, on the fact that continuing to degrade the signal hindered international adoption of GPS, and on impending competition from Russia's nearly complete GLONASS system, which lacked the security feature, giving users a more accurate signal.

The RAND study took a more cautious approach, concluding that commercial GPS growth depended more on declining prices than better accuracy. It argued against turning off Selective Availability until the Pentagon was certain the military had developed effective countermeasures both to deny civilian GPS and GLONASS signals to adversaries during hostilities and to ensure its own ability to acquire and use the military signal in an environment where signals are being jammed. Even if were turned off during peacetime, RAND recommended retaining Selective Availability as a wartime option.

Both studies acknowledged that differential augmentation techniques were rapidly eroding the rationale for Selective Availability and recommended that the Department of Defense shift its focus to better offensive and defensive electronic countermeasures. The military needed to end its reliance on civilian receivers (a holdover from Desert Storm) and develop new receivers that could lock onto the encrypted military signal without first using the civilian signal to acquire satellites, a shortcoming that would prevent jamming the civilian signal to deny it to adversaries. The NRC also recommended adding a second civilian signal, expanding and upgrading the monitoring stations and creating a backup master control station.

When President Clinton issued Presidential Decision Directive, National Science and Technology Council 6 (PDD NSTC-6) on March 28, 1996, it combined elements of both studies. The directive created a "permanent interagency GPS Executive Board, jointly chaired by the Departments of Defense and Transportation" and committed the nation to providing GPS "for peaceful civil, commercial and scientific use on a continuous worldwide basis, free of direct user fees."[72] It charged the Defense Department with continuing to "acquire, operate, and maintain" GPS, designated the Transportation Department as the lead agency "for all federal civil GPS matters," and charged the Department of State with coordinating bilateral or multilateral agreements involving GPS. The National Command Authorities (president and secretary of defense) retained ultimate control of GPS and any government augmentation systems. Most notably, the directive announced the intention to discontinue Selective Availability within a decade and committed the president, beginning in 2000, to make an annual determination about the continued need for it based on input from the various agencies.

At a press conference the next day Vice President Al Gore compared the government-sponsored origin and private sector potential of GPS to that of the Internet. He promised one hundred thousand new high-tech jobs soon and, eventually, the development of handheld phones that could provide precise location information and allow civilian "rescues" similar to Scott O'Grady's. Gore went further than the official text of the president's directive, telling the audience to expect a more accurate civilian signal in "four to ten years."[73] His time frame for the demise of Selective Availability proved accurate, and his prognostication about GPS-enabled phones with navigation capabilities was correct, although that advance was more than a decade away.

Four months after the president issued the directive the FAA awarded a $475 million contract for WAAS to a consortium led by Wilcox Electric, a Northrop subsidiary, but the project soon gained a reputation as a troubled program.[74] Within a year the FAA fired Wilcox and gave the contract to Wilcox subcontractor Hughes Aircraft (acquired in 1998 by Raytheon).[75] By late 1997 the program was under fire in Congress. *Business & Commercial Aviation* reporter Perry Bradley summed up the situation this way: "WAAS has all the hallmarks of a major FAA program: sweeping scope, high technology, contract disputes, allegations of mismanagement, questions about capabilities, congressional posturing, a slipping schedule and a growing budget."[76] Development continued, but the goal of GPS as a "sole means" navigation system for all phases of flight through landings gradually succumbed to concerns about signal interference, jamming, and the need for backup systems. Many ground-based beacons remain today. When the FAA activated WAAS on July 10, 2003, it had spent nearly $3 billion on a system that took so long to construct, some of the regional airlines questioned whether newer technologies might eclipse it.[77] The FAA soldiered on. Five years later the number of certified, published WAAS approaches, 1,333, surpassed traditional instrument landing approaches, and the number reached 2,300 in 2010.[78] WAAS remains a cornerstone of the FAA's "NextGen" overhaul of the National Airspace System, aimed at using technology to save fuel and time, accommodate more traffic, and improve safety. WAAS also provides enhanced GPS accuracy, availability, and integrity for a wide variety of nonaviation users in agriculture, surveying, and surface transportation and is a key example of national infrastructure enabled by and dependent on GPS.

In 1999 the Department of Transportation entered a multiagency partnership with the Coast Guard to add an inland component to the maritime DGPS network, creating the Nationwide DGPS (NDGPS) program.[79] Apart from early criticism that it was redundant with WAAS, this expansion avoided the controversy of its FAA cousin.[80] More than eighty NDGPS radio-beacon sites cover 92 percent of the continental United States, providing enhanced navigational accuracy for surface transportation in addition to maritime needs.[81]

Expanding Consumer Markets

President Clinton's directive included an explicit goal to "encourage private sector investment in and use of U.S. GPS technologies and services." Because GPS originated in the U.S. military, defense contractors and domestic manufacturers had a head start in the marketplace. However, when eight Japanese companies established the Japan GPS Council in 1992, the group grew to seventy-two corporate members by the end of that year.[82] Estimates put the Japanese market at about half the size of the U.S. market, but it was growing fast, particularly the car navigation segment.[83]

Among U.S. manufacturers, Rockwell International, the key government contractor from the outset of the GPS program, saw the commercial portion of its GPS technology sales rise to 25 percent by 1994.[84] In 1997 Magellan Systems, which created the first commercial handheld GPS receiver in 1989 and supplied the NAV 1000 to Desert Storm troops, bought Rockwell's vehicle navigation business. Eight thousand Hertz rental cars already featured Rockwell's PathMaster GPS system (under the "NeverLost" label), and the company marketed it widely to other rental fleets, public safety agencies, utility companies, real estate and sales professionals, and private vehicle owners.[85] The purchase positioned Magellan as a leader in the worldwide vehicle navigation market, projected to grow from $1.1 billion to more than $3 billion by 2000.[86] Magellan, founded in 1986 in San Dimas, California, as a privately held company, prospered during and after the Persian Gulf War and enjoyed such strong sales in Europe that it established a wholly owned subsidiary in England in December 1991.[87] Magellan made GPS products for commercial boating, outdoor recreation, the military, surveying, oil and gas exploration, and forestry and made components for original equipment manufacturers of systems such as autopilots and vehicle tracking.[88] The company proved attractive to investors. Orbital Sciences, a Virginia-based rocket and space systems manufacturer, bought Magellan in 1994 for $50 million and sold it to Thales Navigation of France for $70 million in 2001. A private equity fund, Shah Capital Partners, acquired Thales in 2006 for $170 million and sold Magellan's Consumer Products Division to MiTAC International, a corporation listed on the Taiwan Stock Exchange, for an undisclosed amount in 2008. Throughout, Magellan, headquartered in Santa Clara, California, has remained a top consumer brand recognized for low-cost GPS units for automobiles and recreational use.

Trimble Navigation, the Silicon Valley supplier of those "green position locators" to soldiers, emerged from Desert Storm a big winner. Charles Trimble, a Hewlett Packard executive, and two partners from Hewlett Packard founded the company in 1978, the same year the first Block I GPS satellite launched.[89] Their original focus was high-end navigation systems for boating using Loran, a ground-based radio-navigation system primarily for maritime users. As the GPS constellation grew and the luxury yacht market declined in the mid- to late 1980s, Trimble shifted his business to GPS products for scientific research, surveying, and other commercial applications. Trimble sold three million shares of stock on the NASDAQ exchange in 1990, becoming the first publicly traded GPS company.[90] From 1990 to 1991 Trimble's sales more than doubled, to $151 million from $63 million, and profits more than tripled, to $7 million from $2 million.[91] Revenues came back down to earth following the war, but Trimble continued diversifying its products and customer base and became a leader in developing real-time kinematics (continuously updated GPS while moving), "plug and play" GPS sensors for laptops or digital devices, and GPS integrated with cellular communications circuitry and other manufacturer's central processing units. Early on Trimble expanded GPS technology applications through partnerships with companies like Caterpillar and Nikon, but after 2003 it increasingly acquired smaller companies to broaden its reach across diverse markets. The company has focused on integrated solutions rather than boxed products. By 2012 Trimble's revenues exceeded $1.2 billion, it had extensive international operations, and it offered more than five hundred products for use in agriculture, construction, communications, engineering, mapping, surveying, and precise timing.[92]

In the United States GPS car navigation remained confined mostly to professional users and private luxury automobile owners until after 2000. By that time, improved signal accuracy had attracted more players to the industry, and declining prices finally opened a mass market for standalone personal navigation devices sold through consumer electronics retailers. The era of suction-cup-mounted GPS units with cigarette-lighter cords had arrived. Leading makers of in-dash car audio systems jumped into the fray, along with companies rooted in GPS. The Consumer Electronics Association reported $32.8 million in U.S. sales of aftermarket navigation products during the first three quarters of 2003, a 33 percent increase over the same period in 2002.[93] Magellan introduced one of the first 3.5-inch touch screen units with built-in street-level maps, points of interest, and turn-by-turn directions—at a retail price of $1,299.[94]

In 2004 "mobile navigation" made its first appearance as a separate heading in NPD Group's "Market Share Reports by Category," an annual report compiled by the Port Washington, New York, market research company.[95] NPD uses point-of-sale data from selected retailers to track a broad array of consumer electronics. The company put the U.S. mobile navigation category total at $72.8 million and reported that the top five ranked brands—Magellan, Garmin, Pioneer, Alpine, and Kenwood—captured 95.4 percent of the market.[96]

Prices for portable car navigation devices dipped below $500 in 2004, while the industry shipped an estimated sixty thousand in-car units, more than doubling the previous year's shipments.[97] Also in 2004 the first nationwide traffic alerts system to guide commuters around tie-ups arrived via XM satellite radio's NavTraffic service, incorporated into Pioneer and Alpine units.[98] Mobile navigation sales rose in 2005 to $114.9 million, and the top five brands continued to hold 95 percent of the market.[99] However, Garmin moved ahead of Magellan to claim the top ranking, and a new brand joined the list, which now read Garmin, Magellan, Pioneer, TomTom, and Alpine.[100] While Garmin boasted the largest market share in the United States, Amsterdam-based TomTom, with the largest share of the European market, claimed the worldwide lead.

A quick overview of these two GPS heavyweights is in order.

Garmin's founders, Gary L. Burrell and Dr. Min H. Kao, formed the company in 1989 to take advantage of the revolution they saw GPS bringing to navigation.[101] Both had worked for major technology companies, including Allied Signal, where Kao led development of the first GPS navigator certified by the FAA.[102] When the company went public in December 2000, it had worldwide annual revenues of $345 million, production facilities in Kansas and Taiwan, and extensive product lines serving the aviation, automotive, marine, and recreational markets.[103] After claiming the top U.S. market share in PNDs in 2005, Garmin remained the leader through 2012, when it reported $2.72 billion in total revenue and sold more than fifteen million units worldwide.[104] However, that was 22 percent below its best year, 2008, when it achieved $3.49 billion in revenue.[105]

TomTom, founded in 1991, initially focused on developing software for mobile devices such as handheld barcode scanners for businesses and personal digital assistants (PDAs) for individuals.[106] It launched its first navigation application for PDAs in 2002 and its first standalone PND in 2004. PND sales jumped from a quarter-million units in 2004 to 1.7 million in 2005, when the company went public. It then acquired a German company, Datafactory, which specialized in vehicle telematics (information technology across long distances), giving TomTom entrée into fleet management services. Two years later TomTom acquired an automotive engineering division of Siemens and one of the two largest global mapping companies, Tele Atlas, to position itself for entry into the in-dash navigation market. TomTom reported €1.1 billion in total revenue in 2012, a 17 percent annual decline due to lower PND sales and 36 percent below its best year, 2007, when it achieved €1.74 billion in revenue.

By 2006 the top three brands—Garmin, TomTom, and Magellan—commanded 85 percent of the U.S market, with Garmin alone accounting for nearly half.[107] New suppliers and new products poured into the marketplace, with entrants like long-established mapmaker Rand McNally hoping to capture part of an estimated three-million-unit North American market.[108] The number of PND units shipped in 2006 increased nearly seven-fold over the prior year, while total revenues grew 262 percent, despite falling prices.[109] At the annual Consumer Electronics Show in January 2008, low-end units started at $199 or less, and vendors sought to maintain their profit margins and differentiate their new PNDs by offering larger screens, voice input of addresses, and real-time traffic and weather options.[110] Final 2007 figures showed North American shipments grew to 8 million units and worldwide units shipped reached 34.8 million.[111] Garmin, maintaining its hold on 47 percent of the U.S. market and doubling its European shipments, edged past TomTom in 2007 as the world leader, according to Canalys, a market intelligence consultant to the high-tech and telecom industry.[112]

During 2008 and 2009 two forces converged to disrupt skyrocketing PND sales—a financial panic that caused a deep recession and the rise of inexpensive or free GPS applications for smartphones. The prospect of smartphones as competitors to PNDs had loomed on the horizon for some time, but technical limitations kept phones from delivering the same functionality. Cell phones began to get "smarter" around 1999 as manufacturers found ways to combine them with PDAs. Integrating contact lists and meeting schedules into the phone itself was a natural marriage, since the same customers used both and disliked carrying separate devices. Motorola was first that year with a credit-card-size device called the Clip-On Organizer, which snapped onto the back of the StarTAC phone, enabling automatic dialing and syncing via a cable to a personal computer.[113] As carriers improved data transmission capabilities to handle mobile e-mail volume, using phones to access the Internet became more common. Search engine leader Google began offering its Google Local service—Internet searches tailored to nearby points of interest—to Java-enhanced cell phones in 2005, but screen sizes remained a drawback. Trying to hold a 1-by-1.5-inch screen close enough to read a map while driving was "a moving violation in the making," observed reviewer Dawn C. Chmielewski in the *San Jose Mercury News*.[114] Screen sizes and Internet connectivity continued to grow. Apple introduced the iPhone on June 29, 2007, in an exclusive deal with AT&T that limited distribution to Apple's stores and the telecom's 1,800 retail outlets and direct mail operations.[115] Consumers nevertheless purchased more than 1.3 million iPhones over the next three months, making it AT&T's top-selling phone and the fourth-best selling in the industry.[116] In July 2008 Apple launched the iPhone 3G and a software upgrade enabling real-time GPS mapping and navigation through applications downloaded from its online App Store. Apple soon claimed the number-two spot behind Research in Motion, maker of the Blackberry, and smartphone revenues increased by 71 percent between July 2007 and July 2008 even as overall cellular revenues declined.[117] Analysts estimated that smartphones accounted for 19 percent of all handset sales in 2008, compared to 9 percent the year before.[118] Third-party applications for smartphones, including GPS navigation apps from companies such as Garmin and TomTom, proliferated. Industry watcher iSuppli counted more than three thousand navigation-related apps in Apple's App Store by mid-2009.[119] Inrix Traffic, a free app introduced in August 2009, turned its users into data generators by tracking the location and speed of their phones to create the first "crowdsourced" traffic reports.[120] GPS capability in smartphones enabled users not only to navigate from one point to another but also to receive notices that a friend, favored retailer, or other point of interest was in the vicinity. The term *location-based service*, or LBS, entered the mainstream lexicon, and an ABI Research analyst described the segment as experiencing "nothing short of a gold rush."[121]

Google meanwhile promoted its Android software platform to third-party developers. The Android market by late 2009 offered about twelve thousand apps, compared to Apple's then roughly one hundred thousand apps.[122] With the iPhone (still exclusive to AT&T) gaining market share and luring away existing Verizon customers, competitors fought back. In October 2009 Google unveiled free voice-prompted, turn-by-turn navigation for Android users. For Google, offering free navigation was a way to increase mobile search traffic; phone manufacturers and wireless carriers saw an opportunity to sell more phones. Motorola and Verizon immediately announced the Droid, the first Android-based smartphone to run on the biggest U.S. wireless carrier. The news sent shares of Garmin and TomTom tumbling as analysts considered the implications for PNDs.[123] Three months later Finland's Nokia, the world's largest mobile phone manufacturer, began bundling free navigation with its phones as part of a strategy to compete with Apple's iPhone. Nokia had spent $8.1 billion the previous year acquiring digital mapmaker Navteq and had sold about eighty-three million smartphones with GPS to that point.[124] An automotive industry analyst called it "another nail in the coffin for PND makers."[125] While diversified PND makers survived, PNDs today are a much smaller product category—a market that ABI Research predicts will decline by 40 percent by 2016.[126]

The rise of smartphones signaled a new era of "connected cars," in which drivers can transfer into vehicles the online services available on their phones—a large portion of which require GPS to provide location-based results. By 2010 more than 80 percent of mobile phones and nearly 100 percent of smartphones sold worldwide had Bluetooth wireless connectivity.[127] In the United States 93 percent of model year 2010 vehicles offered Bluetooth as a standard or optional accessory, and 75 percent of models in Western Europe offered it.[128] By mid-2011 about 72.5 million U.S. consumers owned smartphones, and the seven hundred million smartphones in use worldwide represented about 15 percent of all mobile phones.[129] The market research firm Forrester predicted that by 2013 mobile phones would become the most common method of personal navigation as people relied on the device they always have with them—their phone. That prediction was too conservative; the number of navigation-enabled mobile phones doubled in 2011 to an estimated 130 million worldwide, putting them in striking distance of the 150 million PNDs then in use.[130] This represents a massive change over less than a decade in the way a significant portion of all people find their way around in daily life. So, how did phones get GPS capability in the first place?

Civilian Search and Rescue

With mobile phones came the first wireless 911 calls, which proved invaluable for emergencies like automobile accidents, where landline telephones might be far away. By 1994 U.S. cellular networks handled about a half-million 911 calls each month, and the number of cell phone users was growing 48 percent annually.[131] As cell phones and wireless 911 calls proliferated, problems became apparent. Unlike the situation with wired phones at fixed addresses, 911 call centers could not automatically identify a wireless caller's number and location. And unlike downed pilot Scott O'Grady, who knew his location but could not reach rescuers with his radio, an estimated one-quarter of mobile 911 callers who reached an operator could not provide their location—hikers lost in the woods, for instance, or injured car crash victims unable to speak.[132] Worse, callers sometimes could not even connect to a 911 operator if they were outside their own cellular network.

Public safety organizations urged the Federal Communications Commission to require that wireless providers install technology to identify the location of a mobile 911 caller. They worried that increasing cell phone use would render call-center infrastructure obsolete, putting lives and property at greater risk. Wireless industry groups questioned the technical feasibility and called mandates and compliance dates "undue burdens" that would hurt their industry and the economy.[133] Estimates of the cost to implement the enhanced 911, or E-911, requirements ranged from $550 million to $7 billion.[134] In June 1996, after forging an agreement between public safety organizations and the Cellular Telecommunications Industry Association (CTIA), the FCC gave all wireless carriers one year to begin transmitting out-of-network 911 calls and a year and a half to provide 911 dispatchers with the wireless caller's phone number and nearest cell tower.[135] Within five years wireless carriers were required to develop ways to give emergency operators the caller's location within a roughly four-hundred-foot radius at least two-thirds of the time.[136] To cover upgrades at local 911 call centers, most states levied taxes—the familiar E-911 surcharge that appears on phone bills.

Two main alternatives emerged for complying with the E-911 mandate; both were early-stage technologies far from being ready to deploy. One method, known as radio triangulation, calculated a caller's position based on the slightly different times the phone's signal arrived at three network antennas. The other approach required adding GPS receivers to individual handsets. The triangulation method concentrated costs on the network infrastructure, while the GPS solution shifted most of the costs to the handset.[137] Triangulation worked better in "urban canyons," where tall buildings block GPS signals; GPS worked better in rural areas with fewer cellular antennas. However, GPS posed an additional drain on the phone's battery and required more time to phase new phones into the subscriber base.[138]

By this point GPS receivers had been shrunk to postage-stamp-size integrated circuits—silicon chipsets combining radio frequency receiver functions and central processing units for GPS signals. GPS chipmakers embraced the E-911 mandate as an opportunity to sell their technology to phone manufacturers. SiRF Technology Holdings, founded in 1995 by a former Intel executive and based in Santa Clara, California, announced soon after the FCC decision that it had developed advanced GPS chipsets that could more quickly acquire even weak satellite signals while filtering out error-causing multipath (reflected) signals.[139] SiRF turned its attention to reducing power consumption and within two years found a way to cut it significantly, enhancing GPS suitability for mobile devices.[140] In what one industry analyst called "a huge coup" for the three-year-old company, on August 10, 1998, SiRF announced deals with wireless market leader Nokia, phone maker Ericsson of Sweden, and the American division of Japanese microelectronics giant Hitachi.[141] Nokia invested $3 million in the company, while Ericsson and Hitachi licensed SiRF's technology.[142] SiRF became for a time the largest supplier of GPS chips for computers, PNDs, and cell phones. Other competitors joined the action, targeting what became a $270 million market for GPS integrated circuits by 2000, forecast at the time to grow to $1.2 billion by 2005.[143] Qualcomm, a San Diego–based phone maker that was shifting its focus to wireless semiconductors, purchased SnapTrack, a SiRF rival, in January 2000 for $1 billion in stock and within a few years became a serious challenger to Nokia in the worldwide market for wireless chips.[144] SnapTrack developed a hybrid location method that combined GPS in the handset with proprietary software in network computer servers. Other companies, including SiRF, Motorola, and Global Locate, developed variants of this "assisted GPS" approach in which wireless network servers send signals to the phone that compensate for weak or lost GPS signals indoors.[145]

The expansion of the privately held Global Locate illustrates how smaller GPS developers could rapidly grow through licensing. In 2004 Global Locate licensed its GPS chip technology to Infineon Technologies of Germany for a small fee per chip, and over the next five years Infineon sold eighty million of the chips worldwide, including those in the iPhone 3G.[146] As the GPS semiconductor industry matured, the integration of GPS with other functions, such as Wi-Fi and Bluetooth, on a single chip became the norm.[147] This precipitated the larger dominant semiconductor players acquiring the GPS-only chipmakers. Broadcom, a diversified public company based in Irvine, California, with a $15 billion market capitalization, bought Global Locate in 2007 for $226 million.[148] UK-based Bluetooth maker CSR bought SiRF in 2009 for $136 million.[149] By 2010 Broadcom overtook CSR/SiRF as the leading GPS chip supplier after Apple chose its Global Locate–based chips (no longer licensed to Infineon) for the iPhone 4 and iPad.[150] In 2012 Korean electronics giant Samsung, a major Apple rival, bought CSR's handset connectivity and GPS semiconductor division for $310 million. Today GPS, used for both location and precise timing, is literally at the heart of the integrated circuitry powering many consumer electronic devices, and the design and manufacture of GPS chipsets is a strategic battleground in the broader electronics industry.[151]

Meeting the FCC's E-911 mandate proved to be a complicated affair—one with moving targets and deadlines. It became clear that wireless carriers could not comply with the original 2001 deadline. Network-based solutions still lacked the required accuracy; handset-based solutions required more time to get new phones into customers' hands. After SnapTrack announced in 1999 that its GPS technology could locate wireless callers to within 160 feet, the FCC gave wireless carriers the option of meeting that new accuracy level by 2003 or choosing a network solution with a 330-foot accuracy level by 2001.[152] When October 1, 2001, arrived industry-wide progress lagged so much that the FCC had little choice but to issue waivers. The agency extended the 2003 deadline to December 31, 2005, but raised the required portion of subscriber handsets with tracking ability to 95 percent. By 2006 most carriers that chose network-based solutions, such as Cingular and T-Mobile, were in compliance but those that chose the GPS-handset approach, including Verizon, were not because too few customers had upgraded their phones.[153] Mergers brought more complications. Sprint was in compliance until it acquired Nextel, whose customer base had older phones.[154] Industry analysts attributed wireless carriers' intense marketing efforts touting new features and offering rebates and prizes as partly driven by efforts to meet the FCC mandate.[155] The promotions raised consumer awareness of the benefits that GPS and location-based services brought to mobile phones and sparked entrepreneurs to envision new applications. In any case, consumers have the FCC and public safety agencies to thank in part for the amazing array of GPS apps that are available for their smartphones—as well as for the fact that their daily movements may now be tracked via the same technology (see chapter 9). As the first decade of the twenty-first century ended, Arthur C. Clarke's 1956 vision of a satellite-based worldwide network where "no-one on the planet need ever get lost or become out of touch with the community" was reality.

Time for Anxiety

One further illustration of how GPS evolved hand in hand with the digital revolution seems amusing in retrospect, but the anxiety was real at the time. As the year 2000 approached fears spread about the so-called millennium bug or Y2K computer problem—system failures caused by software with the year coded in only two digits. Computer programmers shortened years to two digits in the 1950s, when digital memory was scarce and expensive. The custom persisted. Unknown effects of such programs rolling over from "99" to "00," making 2000 indistinguishable from 1900, threatened not only mainframe networks but also personal computers and countless devices with embedded computer chips. Predictions ranged from the inconvenient—stalled cars, stuck elevators, and power blackouts—to the apocalyptic—chaos in the streets, planes falling from the sky, and nuclear missiles launching themselves.[156] Governments and corporations spent more than $250 billion worldwide on software and hardware fixes in the last few years leading up to the rollover.[157]

In this charged environment, news reports trumpeted a similar but unrelated potential problem for GPS receivers at midnight on August 21–22, 1999. "Coming Satellite Adjustment a Y2K 'Dress Rehearsal,'" announced the *Globe and Mail* (Toronto, Canada).[158] "GPS Users May Want to Dust Off Compasses," advised the *Washington Times*.[159] "Date Flaw Could Sink Navigation Systems," warned *Computing* magazine.[160] However, the GPS "end-of-week" rollover was neither a flaw nor unexpected; it is an integral part of the way GPS marks time. GPS does not operate on a 365-day year. The largest unit of time in the GPS system is one week, or 604,800 seconds.[161] Starting at week zero, the first one pegged initially to midnight (Greenwich mean time), January 5–6, 1980, the system operates for 1,024 weeks, about twenty years, and then rolls over like an odometer to week zero again. This provides a continuous time scale, unlike Coordinated Universal Time (UTC), maintained by the U.S. Naval Observatory, which periodically adds leap seconds to adjust for the earth's rotation in the same way that leap years adjust the calendar for the earth's orbit. There is nothing significant about the 1,024-week cycle. It results from the scheme used to represent each week in ten-digit binary code—zeros and ones—in the stream of digital signals GPS satellites transmit to receivers.[162] In a ten-digit sequence the number of unique combinations of zeroes and ones is 1,024—two raised to the tenth power (2^{10}).

The rollover had no effect on the GPS satellites themselves but put at risk receivers built before 1993, when manufacturers adopted new specifications designed to accommodate it.[163] Nevertheless, GPS hardware makers, which analysts estimated to make up a $4.4 billion industry in 1998, perceived significant liability issues and issued public warnings.[164] Trimble Navigation warned consumers on its website to avoid using GPS devices during the week before the rollover, calling such use "potentially hazardous."[165] Federal agencies from the Department of Transportation to the Consumer Product Safety Commission issued advisories referring consumers to the U.S. Coast Guard Navigation Center's website or a toll-free Y2K hotline (888-USA-4Y2K, later acquired by a savvy telemarketer) for inquiries about devices or manufacturers.[166] Canadian government agencies also launched awareness campaigns, and on the eve of the rollover, the Canadian Coast Guard had sixty search-and-rescue vessels on standby to help stranded boaters.[167] Dar Moja, the only GPS seller in Saudi Arabia, offered free software upgrades to the five thousand customers in the kingdom it had identified as having older units.[168]

Ultimately the rollover caused no serious problems. A spokesman at the U.S. Air Force Y2K Fusion Center at Maxwell Air Force Base in Alabama termed it a "non-event" and reported fewer than twenty calls.[169] A Coast Guard spokesman reported less than a dozen malfunctions among several thousand receivers on its boats, aircraft, and vehicles, and all but one of those reset itself.[170] Rockwell Collins received no reports of malfunctions among the two hundred thousand military units in the field.[171] One Australian naval patrol boat crew saw their GPS receiver fail.[172] The most trouble occurred in Japan, where consumers were early adopters of electronic products. Thousands of complaints poured in about blank or frozen car navigation screens, and estimates put the number of device failures at one hundred thousand.[173] Pioneer Electronics alone sold 270,000 navigation units between 1992 and 1996, but about a quarter of them were not replaced or upgraded in time.[174] American drivers started buying automobile navigation systems later, and Ford and General Motors officials confirmed that there were no problems with the newer equipment.[175]

Four months after the GPS and Y2K rollover worries passed came the official act that GPS manufacturers and users had long awaited—President Clinton signed an order permanently turning off Selective Availability on May 1, 2000.[176] In announcing the decision he noted that modernization plans included adding two new civilian signals as replacement satellites were launched, and he reiterated the commitment to provide GPS free of charge. If the 1996 presidential directive announcing the government's intention to take these steps acted as a sort of starting gun in the race to commercialize GPS, the fulfillment of signals up to ten times more accurate shifted private development into high gear, as shown by the amazing growth of the GPS industry over the first decade of the new millennium.

However, the government did not hand off GPS to private industry like a baton. More than ever, it remained a critical military system, and new security concerns arose sixteen months later with the September 11, 2001, attacks on the World Trade Center and Pentagon. GPS returned to the battlefield in Afghanistan and Iraq and on the home front found unexpected uses and familiar questions.

Firefighters and crews at Ground Zero began the search, recovery, and cleanup effort by dividing the sixteen-acre site into seventy-five-foot-square grids but soon realized that manually mapping the myriad fragments of human remains and other evidence would take too long and yield too many errors.[177] They switched to a system that used handheld computers and bar-coded tags, cataloguing each item as it was discovered with a description, precise time stamp, and exact location via GPS.[178] Original estimates of the cost of hauling 1.8 million tons of wreckage from Ground Zero to Fresh Kills Landfill on Staten Island ran as high as $7 billion.[179] By attaching GPS trackers to about two hundred dump trucks and five boats, officials optimized traffic routes, prevented theft, and increased efficiency by 150 percent—raising loads per vehicle from four per day to ten, cutting the number of trucks, and lowering the final hauling cost to about $750 million.[180]

Troops in Afghanistan found a novel use for GPS that Col. John Mulholland, commander of the 5th Special Forces Group, said helped connect their mission there to the events back home. A fellow officer with family ties to New York procured a strip of metal from one of the towers. The troops cut it into small pieces and buried one in each area where they operated, using GPS to log an exact location, which commanders tracked on a map.[181]

Within weeks of the attacks Americans learned that the Federal Bureau of Investigation was focused on the terrorists' use of handheld GPS devices. That could explain how inexperienced pilots navigated jumbo jets to targets many miles off assigned routes, a task one federal official characterized as more challenging than flying the planes.[182] In November 2001 the captain of a major U.S. carrier reported testing whether he could pick up satellite signals on a handheld device in his cockpit, but the heated windshield and windows blocked them.[183] Others pointed out that the clear skies on September 11 made the Twin Towers visible for miles and the hijackers of American Airlines Flight 11 had only to follow the Hudson River to New York City.[184] It is unclear whether the terrorists possessed the expertise to reprogram the aircraft navigation systems based on coordinates stored in handheld units. The owner of an aviation supply business who sold a GPS handset to one of the hijackers speculated in a television interview that the terrorists could have visited the buildings and set the devices in advance to ensure their accuracy.[185] Numerous print and broadcast reports, including ones on ABC, NBC, and CNN, appeared in May 2002, citing unnamed federal sources suggesting that Mohammed Atta, who piloted Flight 11 into the North Tower, cased the World Trade Center site a day or two before the attack to capture the coordinates on a GPS unit.[186] While the official timeline of his travels makes this theory problematic, it does not rule out the use of portable GPS receivers at another time by him or others for that purpose.

Authorities reconstructed the movements and activities of all nineteen hijackers and those of Zacarias Moussaoui, the so-called twentieth hijacker. He was arrested on immigration charges August 17, 2001, after instructors at a flight school in Minnesota became suspicious. Moussaoui proved to be a key to unraveling the wider plot, and he was later tried, convicted, and sentenced to life in prison. The FBI pieced together financial transactions, physical evidence, and eyewitness accounts confirming multiple purchases of GPS units, cockpit instrument diagrams, operating manuals, maps, video simulations of jetliner operations, and varied simulator and airborne flight lessons. This evidence figured prominently in the Moussaoui indictment and trial.[187] Prosecutors revealed that he had emailed both Garmin and Magellan to ask whether he could convert a street navigation unit for use in a plane.[188]

This type of publicity produced anxiety about GPS. *U.S. News & World Report* titled an article about the potential for good and evil uses of GPS "A Jekyll and Hyde System."[189] Flying airplanes into buildings seemed an unexpected fulfillment of military officials' worries about a "poor man's cruise missile" a decade before. Many expected the government to reactivate Selective Availability; others suggested restricting GPS receiver sales, but a *Forbes* writer likened that to putting toothpaste back in the tube.[190] Analysts calculated that the civilian GPS market had grown to $14 billion a year and could reach $20 billion by 2004.[191] Most experts saw the path forward in terms of exercising greater dexterity in using the technology. Some envisioned programming "virtual no-fly zones" into flight management systems to prevent hijackers—or regular pilots, for that matter—from flying where a plane should not go.[192] President George W. Bush, in a September 27, 2001, speech at Chicago's O'Hare Airport, indicated the government was studying ways for air traffic controllers to override cockpit controls and land planes by remote control when necessary.[193] Pilots resisted such approaches, and tighter security coupled with hardened cockpit doors addressed those concerns. However, remotely controlled aircraft soon appeared in the skies over Afghanistan in the form of unmanned Predator drones, and the use of drones has proliferated.

The 9/11 attacks prompted reexamination of the approach undergirding President Clinton's decision to deactivate Selective Availability—namely, "selective deniability." This means the ability to deny adversaries the use of GPS signals in specific areas where U.S. troops are engaged in operations, without affecting the signals elsewhere. At the same time, U.S. forces or allies must be able to continue using the military signals and to overcome any adversary's attempts to block them through jamming. GPS signals from space are very faint and can easily be jammed using inexpensive equipment. The 9/11 attacks hastened development of sophisticated antijamming techniques and electronic warfare countermeasures.[194]

At 9:25 a.m., September 11, thirty-nine minutes after the first plane struck the North Tower, Ben Sliney, the national operations manager at the FAA's command center in Herndon, Virginia, ordered a full nationwide ground stop. No flights, private or commercial, could take off. It was the first time anyone had given such an order, and it was his first day on the job.[195] Twenty minutes later, after Flight 77 crashed into the Pentagon, Sliney ordered every plane in U.S. airspace—nearly 4,500 of them—to land.[196] Within an hour about three-quarters of them were on the ground, and by 12:16 p.m. the only planes in the air were military aircraft or emergency flights.[197] Controllers diverted hundreds of international flights en route to the United States. The FAA halted civil aviation for two full days and lifted restrictions gradually, starting with airlines and then charter and corporate flights. General aviation restrictions remained until mid-December.[198] Despite these unprecedented circumstances, authorities at no time disrupted the GPS Standard Positioning Service. On September 17, 2001, the Interagency GPS Executive Board (IGEB) posted a statement on its website reaffirming that the U.S. government had "no intent to ever use SA again."[199]

Some nations, however, continued to chafe at their growing dependence on GPS, fearing that the United States could "wreck their economies with the flick of a switch."[200] One month after the IGEB notice, international aviation officials gathered for the ICAO conference in Montreal pressed the United States on unresolved GPS issues, and the Europeans in particular pushed for some type of binding international regime.[201] While Russia's GLONASS system, which attained a full complement of twenty-four satellites in 1995, had deteriorated by 2001 to just six operational satellites, the EU forged ahead with its plans to have thirty Galileo satellites in place by 2008.[202]

This emerging competition for global navigation satellite superiority raised political and security concerns about whether Galileo's signals would interfere with GPS's military bandwidth.[203] After protracted negotiations, with the Iraq War raging in the background, the State Department announced in June 2004 that Secretary of State Colin Powell would sign an agreement with the European Commission covering interoperability between Galileo and GPS. EU analysts at the time estimated that the global satellite navigation hardware and services market had doubled, from $12 billion in 2002 to $24 billion in 2003, and forecast it to reach an astounding $364 billion by 2020.[204]

On December 8, 2004, President Bush signed a presidential directive creating the National Executive Committee for Space-Based Positioning, Navigation and Timing.[205] This successor to the interagency board President Clinton established in 1996 updated the structure and mission to reflect changes in the government, the marketplace, and security needs over the preceding decade. While the secretaries of defense and transportation continued to chair the committee jointly, its membership expanded to include officials from the Departments of State, the Interior, Agriculture, and Commerce, the newly created Homeland Security Department, the Joint Chiefs of Staff, and NASA, as well as observers from the White House and FCC as liaisons. Perhaps most tellingly, the directive established a permanent staff for the organization—a civilian bureaucracy in Washington DC being the surest sign that GPS had become a permanent and integral part of daily life.

Where Are We?
GPS and GNSS Today

I once was lost but now am found.

John Newton, "Amazing Grace," 1779

Like millions of motorists, thirty-two-year-old Jeramie Griffin, a construction worker in Lebanon, Oregon, received a GPS portable navigation device for Christmas in 2009.[1] Black Friday price wars heralded the holiday shopping season with a model from Garmin for eighty-nine dollars, while TomTom undercut its rival with a fifty-nine-dollar loss leader, making its PNDs the most heavily purchased electronics item online that day.[2] What came unexpectedly with Griffin's gift was fifteen minutes of fame. He tried his new GPS unit for the first time Christmas Eve, discovering a shortcut he believed would trim forty minutes off the four-hour trek to his fiancée's relatives, who lived about two hundred miles away, on the other side of the Cascade Range from the couple's home in the Willamette Valley.[3] They packed their silver Dodge Durango, bundled up their eleven-month old daughter, and departed about 3:30 p.m. Their GPS guidance took them east into the mountains on a state highway, then northeast along a local road, and finally onto one of numerous unplowed logging roads that crisscross the national forests, where they got stuck in deep snow.[4] Satellite navigation carried them beyond cell phone coverage. They spent a harrowing night in the car, even making a farewell video.[5] Frantic relatives contacted two close friends, who found the stranded family about twenty-four hours after they drove into the snow bank. The friends borrowed a GPS unit identical to Griffin's and duplicated the route he used.[6] Griffin's family was one of three groups of travelers stranded that holiday weekend after using GPS to plot routes across remote, snow-clogged Oregon roads. They drew the attention of the Associated Press and then CNN.[7] Soon their stories went viral on the Internet. The Air Force felt compelled afterward to issue a statement reminding the public that it operates the satellites that emit GPS signals but neither creates nor updates the maps in devices, nor is it involved in calculating routes between destinations.[8]

Since GPS navigation devices became mass consumer products many have misused them, placed too much faith in them, or blamed them for unexpected outcomes. Stories abound of people driving into swamps or onto railroad tracks because, they say, the GPS unit directed them to do so. One of the first television commercials to tap this trend showed a driver crashing into a storefront as the GPS voice said, "Turn right [pause] in two hundred feet." The popular sitcom *The Office* had its main character robotically chant, "The machine knows," as he turned off a road—and into a lake.[9] Allstate Insurance commercials featured the personified "Mayhem" posing as a GPS unit giving incorrect instructions because "you never update me, so now I just have to wing it." By 2008 more than eight thousand visitors annually showed up in vehicles at the front door to Ireland's famous Stone Age monument Newgrange after entering the landmark into GPS units.[10] Ignoring the facility's published directions in favor of self-guidance, all had to be sent south across the Boyne River to the visitor's center, where official tours originate. British soldiers who repeatedly steered army tanks and gun carriers down a narrow lane in Donnington, England, some fifteen miles from their barracks blamed receiver error, and British railroad officials blamed poor mapping software for tall trucks frequently striking low railroad bridges, causing millions in damages.[11] A Beaumont, Texas, neighborhood near the Port of Orange reported a similar problem in 2009, when big rigs began rolling down and having to back out of Childers Drive, a narrow, tree-lined cul-de-sac. Drivers using GPS units were trying to find Childers Road, an industrial street several miles south lined with warehouses and shipyards.[12] A contractor in Carrollton, Georgia, demolished the wrong house in 2009 after an apparent mix-up involving GPS coordinates.[13] Angry Hollywood Hills residents, exasperated by cars and tour buses clogging narrow roads and dead-end streets near the famous Hollywood sign, finally convinced Google Maps and Garmin in early 2012 to change the programming in their systems and guide users to appropriate viewing sites farther away.[14]

Media reports of GPS users gone astray seem to have waned, perhaps indicating that device makers have corrected most map glitches or that users have become more cautious, but such stories still appear occasionally. Tractor-trailer drivers using GPS units to locate Broadhead Road in a Bethlehem, Pennsylvania, industrial park were still finding themselves stuck on Broadhead Court, a narrow, nineteenth-century lane, a local newspaper reported in February 2012.[15] Days later the *Costa Concordia* cruise ship ran aground and tipped over off the Italian coast. A former cruise industry executive acknowledged that technologies such as underwater sonar and GPS made many crews complacent about safety.[16] The *London Telegraph* reported that buses transporting American and Australian teams from Heathrow Airport to the 2012 Olympic Village got lost, turning a ninety-minute trip into four hours, because the destination was not programmed into the vehicles' GPS systems and the drivers did not know how to operate them.[17] On the first day of competition hundreds of thousands of spectators lined the men's cycling course, using mobile phones to send photos, videos, and text messages. Overwhelmed data networks blocked the information from GPS units traveling with the cyclists, so television commentators could not report how far ahead of the pack the leaders were.[18]

A survey in 2010 showed that more than half of British drivers do not trust GPS devices; a third said they have gotten lost while using them and 15 percent blamed them for making them late to an important event.[19] Half of Australian drivers reported a lack of trust, while two-thirds blamed GPS units for getting them lost and a third acknowledged frustration with using their device.[20] Michelin, which publishes maps and travel guides in addition to manufacturing tires, surveyed 2,200 U.S. adult drivers in April 2013 and found that 63 percent said GPS devices had led them astray at least four times, while 7 percent reported being misdirected more than ten times.[21] Allstate Insurance reported that a study of three thousand drivers found 83 percent of men and 75 percent of women occasionally disregard their turn-by-turn directions, while a third keep a map in the car in case they need one.[22] States still print millions of paper road maps to distribute at highway rest stops and tourist venues, but lower demand and tight finances make such programs attractive targets for budget cutters.[23] Georgia reduced the number it prints from 1.7 million to 1.4 million between 2004 and 2009.[24] Iowa in 2012 cut the number of state road maps it printed from 1.4 million to eight hundred thousand, saving about $240,000.[25] AAA and Rand McNally acknowledge they are printing fewer maps but decline to give numbers.[26] Insurance companies have an interest in public perceptions of GPS reliability. Many are actively promoting GPS tracking devices designed to record driving behavior and reward safer drivers with lower premiums.

The Bell on the Cat

While some drivers lost patience with their navigation devices, many others were losing the units themselves—to thieves. As sales rose, so did car burglaries. The FBI reported that GPS thefts from vehicles jumped 30 percent between 2000 and 2004.[27] Even as thefts of automobiles plunged 13 percent in 2008, car burglaries increased 2 percent.[28] Philadelphia police said GPS units accounted for 20 percent of all thefts from vehicles.[29] Dublin, Ireland, police put the figure at 25 percent.[30] Spanish officials blamed GPS units and cell phones for car burglaries having "skyrocketed."[31] Melbourne, Australia, police recorded more than six hundred GPS units stolen from vehicles every month in 2008.[32] Bethesda, Maryland, police echoed an admonition to car owners that was prominent in GPS-related theft stories: leaving suction-cup mounts on car windshields was a red flag for thieves.[33] Media reports of this kind also seem to have decreased. Drivers may have learned to be more careful. More likely the decline is because GPS navigation has migrated to smartphones, which people tend to take with them when they leave the vehicle, and to a growing number of in-dash systems, which are less prone to "smash and grab" break-ins. Industry analysts predict factory-installed navigation systems will overtake PND shipments by 2015.[34] However, a mid-2012 report highlights the danger of car owners programming their address into GPS systems as "home." A car thief drove his victim's Lexus to her house, used the garage opener to let himself in, burglarized the home, and abandoned the vehicle nearby.[35]

When thieves steal PNDs, the odds of recovery are slim. Stolen or lost smartphones are another matter. The GPS infrastructure by itself cannot track receivers using the signals; tracking requires additional circuitry to transmit the calculated coordinates to some other monitoring system. As discussed in the preceding chapter, that technology is built into phones to meet the FCC's mandate for locating mobile 911 callers, and it has migrated into tablets and some laptop computers. It enables apps that owners of stolen or lost smartphones—

Apple or Android—can access online to display the phone's location on a map. Reports over the past several years from around the world have shown victims and police using these apps to catch phone thieves. A Chicago executive called 911 after two men mugged him and stole his phone, credit cards, cash, and driver's license. He provided real-time location information to the operator, who relayed it to police. The officers found the men at a gas station about a mile away and held them until the victim arrived and identified them.[36] Sometimes such cases go awry. A Sherwood, England, resident was stuck with an $800 repair bill after police broke down his door while responding to an incorrect address provided by phone-tracking software.[37] Such flaws make obtaining a search warrant increasingly difficult, so police often use other tactics. After a man robbed a restaurant delivery driver in suburban Atlanta of his phone, cash, and chicken wings, the driver called police and gave them the phone's location, which was stationary at a residential address. The officers knocked on the door and, while they were conversing with the resident, who matched the robber's description, the victim activated a feature to make the phone ring, precluding the need for a warrant.[38] Antitheft apps are evolving to include features that sound alarms and allow victims to photograph thieves remotely.

The ability to know where someone or something is at all times has produced ambivalent feelings among users and requires a difficult balancing act for policy makers as technology races ahead of social norms and the legal system. On the one hand, GPS creates a multitude of powerful tools for efficiency, safety, convenience, and profit. On the other hand, people worry about the erosion of privacy and misuse of those tools. A person tracking his own movements calls it navigation. When someone else does so without his knowledge, he calls it spying. Those doing the latter type of cell phone tracking primarily call it one of two things—marketing or law enforcement.

GPS tracking devices became big business long before the technology proliferated in mobile phones. The military recognized the logistical advantages of GPS tracking from the outset. As mentioned in chapter 6, Persian Gulf War troops radioed their coordinates to each other both to rendezvous and to avoid fratricide. During the 1990s the military and the private sector began to use electronic tracking devices attached to vehicles for fleet management. By 2004 the Defense Department used GPS to track more than forty-seven thousand commercial-carrier shipments of arms, ammunition, and explosives transported each year across the United States.[39] Today the Army is equipping individual frontline infantrymen with the General Dynamics Rifleman Radio.[40]

Described as a smartphone combined with a tactical radio, it uses GPS tracking to enable a platoon or squad leader to monitor his soldiers' positions on a display attached to his radio.[41]

Municipal transit systems began installing automatic vehicle location (AVL) devices on buses in the early 1990s. Trimble Navigation introduced its Transit-Trak turnkey system on ten demonstration buses shuttling attendees to International Public Transit Expo '90 in Houston.[42] The next year Dallas Area Rapid Transit outfitted 1,600 buses with Washington-based Marcor's Hummingbird trackers, numerically the largest nonmilitary GPS contract awarded to that point and estimated at more than $1 million.[43] Qualcomm, the eventual world leader in AVL devices, had equipped more than a half-million vehicles owned by 1,500 trucking companies by 2004.[44] By 2010 more than two-thirds of all commercial surface vehicle fleet operators in the United States were estimated to have adopted GPS tracking, yielding total economic benefits exceeding $10 billion annually through fuel, labor, and capital savings.[45] AT&T used a phrase from a children's fable to succinctly explain the benefits of fleet tracking in a 2012 commercial. Following an image of a convoy of seven trucks hauling construction equipment down a rural Texas highway, a worker attaches a magnetic GPS tracker to a truck, saying, "This is the bell on the cat."[46] Similarly, 40 percent of rail systems and 75 percent of ferryboat operators had adopted GPS by the end of the decade.[47]

As with other information technology products, prices have dropped along with size and weight, making GPS tracking increasingly affordable and practical not only for smaller fleets but also for valuable assets, for pets, and for people. Trackers weighing less than an ounce and about the size of remote-entry key fobs for automobiles have become loss-prevention tools for items ranging from flat-panel televisions to bank bags. Cambridge, Massachusetts, police caught a bank robber eight minutes after the crime by tracking him to the public bus he was riding. In his pocket they found the stolen $3,600 and a GPS tracker the teller had hidden in the bundles.[48] A Wellington, Florida, community center, fed up with thieves stealing figurines from a nativity scene, attached a GPS tracker to baby Jesus. When it disappeared, police followed the signal and made an arrest.[49] Qualcomm subsidiary Snaptracs offers Tagg, a lightweight, battery-operated tracker that attaches to a pet's collar and sells for $99.95 plus a $7.95 monthly service fee.[50] When pets move beyond a zone or "geo-fence" defined by the owner, Tagg sends a text message or e-mail alert.

People-tracking equipment revenues increased 352 percent between 2005 and 2010, and the personal tracking industry, now growing 40 percent or more annually, will reach $1 billion by 2017, according to ABI Research.[51] Such growth is attracting large companies that already compete in commercial tracking as well as new entrants filling market niches. Various tracking systems now cater to those responsible for infants, children, and people with special needs such as autism or dementia. The Alzheimer's Association collaborated with Omnilink Systems of Alpharetta, Georgia, to create Comfort Zone, a web-based service that families or caregivers can use to monitor Alzheimer's patients.[52] Omnilink, which got its start tracking criminal offenders and now offers numerous commercial solutions, formed a similar partnership with the AmberWatch Foundation, employing the same technology to monitor children.[53] GTX, of Los Angeles, California, collaborated with Teaneck, New Jersey, footwear maker Aetrex Worldwide to embed a tracker in the heel of the Navistar GPS shoe, available for men or women. Each pair costs about $300 plus monthly connection fees, and the company offers a tracking portal on its website.[54] Alzheimer's patients may forget or remove a separate device, but putting on shoes is an aspect of memory most patients retain, say experts.[55] The National Museum of Science and Technology in Stockholm, Sweden, selected the GPS shoe to illustrate the spread of GPS into daily life, including clothing, in an exhibit of "the one hundred most important inventions in history, as rated by the Swedish people."[56] GPS ranked twenty-first at this writing, while computers held first place.

A Swedish company, Purple Scout, makes a bright-yellow safety vest called ChildChecker with built-in GPS tracking that daycare centers there have adopted for outdoor activities and field trips.[57] It reports each child's location to the teacher's smartphone. Administrators and daycare workers praise the device, although some critics wonder about the long-term effects of constant surveillance on children's sense of privacy.[58] The wristwatch phone envisioned in the Dick Tracy comic strip is now available (minus two-way TV) via New York City–based Revolutionary Tracker, which offers its GPS Watch Locator, a multipurpose watch, tracker, and communications device featuring two-way voice and text as well as a listen-in mode. Brightly colored wrist straps increase the appeal for children, or parents can hide the electronic component alone in an infant's clothing or blankets.[59] Amber Alert GPS, located in South Jordan, Utah, offers a personal locator for children with an automatic predator alert that informs parents if their child is within five hundred feet of the registered address of any sex offender listed in the U.S. Justice Department's national database.[60]

The criminal justice system uses GPS technology extensively to track probationers, parolees, pretrial defendants, and people awaiting deportation hearings. Radio frequency ankle bracelets emerged in the late 1970s as a tool to enforce house arrests. Their use expanded when GPS made it possible to track the wearer continuously rather than simply sound an alert when he or she left the house. Supervising officers can monitor travel to work or drug treatment while getting an immediate notification if the offender enters an excluded zone, such as a schoolyard or the neighborhood of a person with a restraining order.[61] The ankle bracelet reached its greatest national prominence, perhaps, in 2005, when the courts sentenced television celebrity Martha Stewart to house arrest on her 150-acre estate.[62] They are not foolproof. The *Guardian* (London) reported in 2011 that a suspect in Rochdale, England, tricked two security guards into attaching a tracking device to his prosthetic leg, which he later removed to violate curfew.[63] When a major electronic monitoring vendor's computer system shut down for twelve hours in 2010, corrections officials in forty-nine states could not follow the movements of sixteen thousand offenders they were tracking, although the offenders were unaware of the lapse.[64] Complaints that correctional systems, in the United States and abroad, rely too heavily on the technology arise whenever offenders commit crimes while under electronic surveillance.[65] A Baltimore woman sued a GPS tracking vendor, Nebraska-based iSecureTrac, in mid-2012 after her five-year-old daughter suffered permanent brain damage from a stray bullet fired by a juvenile offender the company was monitoring.[66] The case is pending. Police rearrested a Massachusetts defendant awaiting trial in early 2012 after a tracking system showed he violated house arrest, but his lawyer successfully argued the readings were false.[67] His expert witness, a civil engineer with extensive GPS experience, testified that the home's construction and the weakness of satellite signals indoors could have affected tracking results.[68]

A 2012 U.S. Justice Department–sponsored study of more than six thousand California sex offenders showed that real-time GPS tracking was 30 percent more expensive than traditional supervision but significantly more effective. GPS monitoring cost about $36 per day per parolee, compared to $27.50 per day for traditional supervision, but traditionally monitored parolees were three times more likely to commit a sex-related violation.[69] Incarceration in California at the time of the study cost about $129 per day.[70]

The American Correctional Association reported that 22,192 people wore electronic monitoring devices in 1999.[71] Today about one in forty-five adults in the United States, more than 4.88 million, is under criminal justice supervision outside a correctional facility.[72] The number of people who wear or carry some type of court-ordered electronic monitor grew fivefold, to one hundred thousand, between 1999 and 2006 and by the end of the decade reached an estimated two hundred thousand.[73] The trend has fueled substantial growth among early players in electronic offender monitoring, attracting multinational public corporations to the business. Colorado-based BI (founded in 1978 as Behavior Interventions), with about a third of the U.S. market, tracks more than sixty thousand offenders through contracts with about nine hundred federal, state, and local agencies in all fifty states.[74] The GEO Group, a $1.6 billion global provider of correctional, detention, and treatment services to all levels of government, acquired BI in 2011 for $415 million.[75] Former U.S. drug czar and Florida governor Bob Martinez cofounded Pro Tech Monitoring in Tampa in 1996.[76] An Israeli company, Dmatek Group, which already owned U.S. monitoring subsidiary Elmo-Tech, bought Pro Tech Monitoring in 2007 for $12.5 million.[77] Dmatek rebranded itself as Attenti in 2010, and then 3M, a diversified worldwide conglomerate with $29.6 billion in sales, acquired Attenti in 2011 for about $230 million.[78] In 2011 the top three offender tracking companies posted combined sales exceeding $61 million. BI posted sales of $26.8 million, and Attenti finished the year with $16.4 million.[79] Relative newcomer SecureAlert, a Utah company founded in 2006, recorded $17.96 million in sales.[80] At least a half-dozen other vendors compete for a share of the market.

Privacy versus Secrecy

As tracking proved useful for those already in the justice system, police stepped up GPS use in criminal investigations. Vehicles impounded after an arrest often have a navigation system from which forensic investigators can tease a stored digital "breadcrumb" trail of its movements.[81] Before an arrest, detectives could attach magnetic trackers to vehicles without alerting a suspect or seeking a judge's permission, and the technology gave police forces, particularly smaller ones with fewer resources, an affordable tool to track suspects, generate evidence, and close cases. For example, police in Fairfax County and Alexandria, Virginia, halted a series of a dozen assaults on women in 2008 after placing a GPS tracker on a van belonging to a suspect living near the crime scenes—a convicted rapist who had served seventeen years in prison. They later caught him dragging a woman from the van into some woods and made an arrest, after which the assaults ended.[82] Civil libertarians and defense lawyers argued that using such tactics without a search warrant are illegal searches and seizures. As cases and appeals involving GPS tracking moved through the judicial system, local courts often sided with police, but state and federal courts differed on the need for a warrant.[83]

In 2011 the first test case involving warrantless GPS tracking, *United States v. Jones*, reached the Supreme Court.[84] FBI agents and metro police had placed a GPS tracker on Washington DC nightclub owner Antoine Jones's Jeep Cherokee and recorded his movements for twenty-eight days while gathering evidence to convict him in a drug-trafficking investigation. During the trial, a government lawyer characterized federal investigations that use GPS tracking as numbering "in the low thousands annually."[85] The aggregate number of investigations involving GPS among all levels of law enforcement across the nation prior to the ruling is unknown, but the decision effectively ended police use of GPS vehicle tracking without a warrant. On January 23, 2012, the justices unanimously ruled that police in the case had violated the Constitution. Justice Antonin Scalia, delivering the main opinion of the court, observed, "We have no doubt that such a physical intrusion would have been considered a 'search' within the meaning of the Fourth Amendment when it was adopted."[86] However, the justices split over the rationale for the ruling. Chief Justice John G. Roberts Jr. and Justices Anthony M. Kennedy, Clarence Thomas, and Sonia Sotomayor agreed with Scalia that the case centered on the physical intrusion created by placing the device on the vehicle. Justice Samuel A. Alito Jr. observed that modern technologies no longer require the traditional "physical trespass" the authors of the Fourth Amendment had in mind. He contended that the issue at stake was the defendant's "reasonable expectation of privacy," a more recent but well-established interpretation of the Fourth Amendment growing out of wiretapping and eavesdropping cases. Justices Ruth Bader Ginsburg, Stephen G. Breyer, and Elena Kagan joined Alito in this view.

Justice Sotomayor, in a concurring opinion, made clear that she agreed with Alito's reasoning, although she was content to dispose of *United States v. Jones* on the narrower grounds Scalia proposed. She suggested that given the volumes of information people share today with retailers, phone companies, and other businesses, protecting societal expectations of privacy would be possible "only if our Fourth Amendment jurisprudence ceases to treat secrecy as a prerequisite for privacy." Police have not traditionally needed a warrant to obtain information voluntarily shared with third parties.

While the case clarified *police* use of GPS tracking, the law remains unsettled regarding *private* uses. State laws vary widely. A Minnesota court in 2011 acquitted a man charged with illegally monitoring his estranged wife with a GPS tracker because he was co-owner of the car.[87] Texas and California prohibit most GPS tracking without consent, but a New York court upheld tracking a government employee in his private vehicle after he had been caught falsifying his time sheets.[88] Cases of this type seem sure to reach the Supreme Court eventually.

Even before the Jones case reached the high court, lawmakers heard growing calls for clarity about how companies and law enforcement agencies can use geolocation data generated by navigation and location-based services in vehicles and smartphones. A media hullabaloo erupted in April 2011 after two researchers announced at Where 2.0, a conference on "the business of location," their discovery that Apple's iPhones and iPads store users' time-stamped location histories in "hidden" files.[89] Apple quickly responded that it does not track iPhone users and explained the files as a "cache" of recent Wi-Fi hot spots and cell tower locations used to speed connectivity, like the temporary Internet files stored in personal computers.[90] Apple characterized the year's worth of location data stored on each phone as a "bug" and promised to fix it with a software update that would limit storage to seven days. The company also acknowledged that it was collecting anonymous traffic data to build a crowd-sourced traffic service like that operated by Google Maps for Android phones, which also cache Wi-Fi and cell tower data. The public uproar prompted a hearing by the Senate Judiciary Subcommittee on Privacy, Technology, and the Law.

By the end of May 2011 House or Senate members had introduced at least ten legislative proposals aimed at various aspects of consumer privacy and data security.[91] A lack of cosponsors and little or no bipartisan support hampered many of them. The leaders of the Bipartisan Congressional Privacy Caucus, Rep. Edward Markey, a Massachusetts Democrat, and Rep. Joe Barton, a Texas Republican, secured more than forty cosponsors, including a half-dozen Republicans, for the Do Not Track Kids Act of 2011, H.R. 1895.[92] It sought to update the Children's Online Privacy Act of 1998 by prohibiting companies from using or providing personal information to third parties about individuals younger than eighteen.[93] Momentum appeared to increase when Rep. Jason Chaffetz, a Utah Republican, and Sen. Ron Wyden, an Oregon Democrat, in June 2011 coauthored the bipartisan and bicameral Geolocational Privacy and Surveillance (GPS) Act, H.R. 2168 and S. 1212.[94] The House bill attracted twenty-five cosponsors, mostly Democrats and some Republicans, but its companion in the Senate gained only one cosponsor, a Republican.[95] Sen. Al Franken, a Minnesota Democrat, simultaneously introduced the Location Privacy Act of 2011, S. 1223, which garnered six cosponsors, five Democrats and one Independent.[96] It was narrower in scope, focusing on commercial service providers, but industry officials in telecommunications, advertising, and marketing cast a wary eye on all such bills.[97] More than a year later, none had emerged from their committees for a vote. Just before the Senate recessed in August 2012, Wyden filed amendments to attach the bipartisan GPS Act to the Cybersecurity Act of 2012, S. 3414, comprehensive legislation aimed at protecting the nation's critical information infrastructure.[98] The Senate ultimately failed to invoke cloture on S. 3414, and the 112th Congress recessed August 3.[99] Clarity on GPS issues would have to wait as Congress faced partisan gridlock, a presidential contest, and uncertain prospects for a lame-duck session filled with even more pressing budget and tax issues.

Chaffetz and Wyden reintroduced the GPS Act, H.R. 1312 and S. 639, in spring 2013 with substantially the same language.[100] As this book went to production both the House and Senate versions remained lodged in their respective committees. GovTrack.us assigned the House bill a 32 percent chance of being enacted, while it gave the Senate version (with only one cosponsor) only a 1 percent chance of enactment.[101] The GPS Act would require the government to show probable cause and get a warrant before acquiring geolocational information about any person in the United States. It would create criminal penalties similar to those for wiretapping for anyone surreptitiously using electronic devices to track a person's movements, and it would prohibit service providers from sharing customers' geolocation information with third parties without customer consent.[102]

Buried in the fine print of a typical provider's terms and conditions is language granting permission to share "aggregated or anonymous information in various formats" with "trusted entities," including retail, marketing, and advertising companies that offer products that may be of interest.[103] That information is valuable to the carriers. Pyramid Research, a Cambridge, Massachusetts-based market research company that follows the telecom industry, projected in mid-2011 that global location-based advertising revenues would reach $6.2 billion by 2015, an exponential rise from $86 million in 2008.[104] Legal restrictions on using personal data could constrain the growth of location-based services, observed Glen Gibbons, editor and publisher of *Inside GNSS* and former editor of *GPS World*, who has followed the GPS industry over several decades. "I expect that location privacy will steadily become more of an issue, in the same way that exploitation of user personal data and behavior by social media has encountered some pushback from citizens generally," Gibbons said. "A key legal question will be whether one's personal real-time location can gain the same status as private communications or whether that is treated as something intrinsically in the public domain."[105]

How much information about an individual does a carrier acquire? A German Green Party politician and privacy advocate wanted to find out, so he sued his provider, Deutsche Telekom (which owns T-Mobile) to gain access to records it had collected about him. In one six-month span the carrier logged his latitude and longitude coordinates 35,831 times.[106] Beyond shaping the content of ads that pop up on a phone when the user is playing an online game with friends, such detailed location information is a magnet for law enforcement officials trying to establish a suspect's whereabouts. Cell phone surveillance has skyrocketed in the past few years as warrants for wiretapping have declined 14 percent.[107] Representative Markey, the cochairman of the Bipartisan Congressional Privacy Caucus, in mid-2012 requested an accounting from nine wireless carriers of requests for subscriber data made by law enforcement agencies at all levels.[108] Reports from eight companies revealed federal, state, and local law enforcement agencies made more than 1.3 million requests for consumer cell phone records in 2011, a figure Markey called "startling."[109] The number understates the total because T-Mobile did not provide a figure, but the company acknowledged that requests had increased between 12 percent and 16 percent annually. Verizon reported a 15 percent annual increase. Requests included "cell tower dumps," in which police trawl through all phone calls connected through a particular tower over a specified period. Cell phone carriers said the volume of requests forced them to hire new technicians and lawyers, and some even outsourced the work to specialists that have sprung up to meet demand.[110] Although carriers may charge law enforcement agencies to recover "reasonable" costs—AT&T collected $8.3 million in 2011—some complained they lost money or that bills went unpaid.[111]

This level of routine police involvement with cell phone records seems certain to spark litigation and legislation. In fact, one California telecom provider took the unusual step in mid-2012 of responding to an FBI request made through a "national security letter" by filing a civil complaint in U.S. District Court.[112] National security letters, under the U.S. Patriot Act, do not require judicial oversight and impose a gag order on the recipient, which in this case means that court documents redact the name of the plaintiff. Since 2000 the FBI's use of such letters has doubled, to more than sixteen thousand annually.[113]

Customers also grant data-sharing rights to the makers of location-based apps they install on their phones. A *Wall Street Journal* investigation found that 47 of the 101 most popular smartphone apps sent location data to third parties.[114] Dutch GPS giant TomTom, searching for revenues to offset slumping device sales, marketed driver behavior data to European governments on the understanding they would use it for transportation planning. The company suffered a public relations black eye when the Dutch newspaper *Algemeen Dagblad* revealed that Netherlands officials shared the data with traffic police, who used it to set speed traps.[115]

Phone hacking is another threat. The rising use of smartphones for personal shopping and banking and their adoption for business and professional use has mobile cybercrime on the rise. Privacy advocates have criticized a type of software called Carrier IQ, sold to mobile carriers to monitor their networks, as spyware because it is capable of logging the user's keystrokes.[116] Whether hackers can access and commandeer the software for other purposes is unknown. Documents and e-mails downloaded from the cloud pose risks of malicious software aimed at everything from identity theft to corporate espionage. Lawyers representing a California software company in a $2.2 billion lawsuit against the Chinese government and several computer manufacturers received nearly a dozen Trojan e-mails (traced to servers in China) that were designed to steal confidential documents from the firm's computer system.[117]

Business users clearly think the benefits of mobile devices and apps outweigh the risks. An AT&T poll in early 2012 found that 96 percent of small businesses used wireless technologies and nearly two-thirds doubted their company's ability to survive without them.[118] Almost a third of small businesses used mobile apps to save time, increase productivity, and reduce costs. Of those who did, four-fifths used GPS navigation and mapping apps and nearly half used location-based services. IE Market Research of Vancouver, British Columbia, forecasts GPS navigation and LBS services to reach $15.2 billion by 2016.[119] TNS Global, a London-based market intelligence consultancy, surveyed forty-eight thousand people in fifty-eight countries and found that 62 percent of mobile phone users who do not already use LBS aspire to.[120] One in five surveyed was interested in seeing mobile advertising if it offered him or her a nearby deal.[121]

Free advertiser-supported LBS apps are likely to be among the first ones any new smartphone owner downloads. Just about any service or subject of interest is more relevant to a user based on location, so news aggregators, sports and weather apps, Internet radio, and many others use GPS location. However, if users open their phone settings to examine the apps listed under location services, they will find some that appear to employ location primarily or only for streaming ads.

Startups and established companies have rushed into the LBS space, offering social networking, customer reviews, shopping guidance, price comparisons, coupons, loyalty programs, and the latest wrinkle—augmented reality. Many apps combine several or all of these elements. LBS apps help people find everyday needs—the lowest-priced gasoline (Gas Buddy, Cheap Gas!, Gas Guru), a clean public restroom (Sit or Squat, Bestroom, Have2P), restaurants (Yelp, Urban Spoon, Open Table), medical care (HealthTap, DoctorsElite, Zoc-Doc), and of course, real estate (Realtor.com, Zillow, Trulia), the industry that coined the phrase "location, location, location." Some newer apps (Highlight, Intro, Sonar) link users through one of the larger social networks, such as Facebook, Twitter, Foursquare, or LinkedIn, which dominate the U.S. market.[122] Social networking is growing faster than other apps. Distimo, a Dutch market research firm that specializes in the mobile marketplace, reported that downloads of the one hundred most popular social networking apps in Apple's App Store for U.S. customers increased 193 percent from mid-2010 to mid-2012, while the store average for all apps was only 43 percent.[123] During that time Facebook remained the world leader, but competitors LINE, WeChat, and Viber surpassed Facebook downloads in many Asian countries from mid-2011 to mid-2012.[124] As of late 2012 both Facebook and Groupon, which have publicly traded shares, were finding it difficult to translate their large mobile presence into earnings that meet investors' expectations.[125]

The LBS market is evolving so rapidly that some well-known names (Gowalla, Brightkite, Babbleville) have already gone under—"deadpooled" is the industry term—or been acquired (such as Loopt, which prepaid card specialist Green Dot will use to enter the mobile wallet business).[126] It seems certain that the mobile app market will undergo extensive experimentation and consolidation over the next several years. Whatever emerges will undoubtedly integrate personal navigation. Navigation apps have evolved far beyond public roadways to include subway systems, bus routes, and large destinations such as Disney World and Arlington National Cemetery. GPS guidance has so altered consumer behavior that Google, Microsoft, Nokia (which owns Navteq), Broadcom, Qualcomm, Research in Motion, and other companies are investing millions to extend accurate navigation—down to the inch—indoors, where GPS signals often fail.[127] Each has patents using a variety of technologies to make this possible. Some use cellular phone signals or Bluetooth beacons at Wi-Fi hotspots; others use inertial methods—gyroscopes, accelerometers, and compasses—to measure movements from a known location, such as GPS coordinates at a building's entrance.[128] Apple has acquired several location technology companies and announced a worldwide agreement with TomTom in mid-July 2012 that suggests it will head indoors as well.[129] Google announced in July 2012 that it had already mapped ten thousand indoor locations, including more than 2.7 million square feet in seventeen Smithsonian museums.[130] That same month Walgreen's announced it had mapped all of its 7,907 stores in a partnership with mapping and search startup Aisle411, a coup that catapulted the small company's indoor map offerings above 9,000.[131] Not only will shoppers know where they are in a store; sellers will know, too. Those Valpak coupons that shoppers are used to getting in their mailboxes will appear instantly on their phones for consumers using the JUNAIO app, an augmented reality browser. The app activates the phone's camera, and geolocation coupons pop up on the screen as the shopper scans his or her surroundings.[132] Augmented reality is attracting such giants as IBM, which is working on an app that would allow shoppers to pan their cameras across an entire grocery aisle and identify products that fit personalized, preselected criteria—and then offer a coupon.[133] Analysts have forecasted the global indoor location market to grow from $449 million in 2013 to $2.6 billion in 2018.[134]

GPS for personal use has come a long way from the early days of geocaching, a type of high-tech treasure hunt where people hide small artifacts and post their latitude and longitude for other hobbyists to track down, but in many ways the future of location-based socializing and shopping for bargains reflects those roots.

Commercial and Professional Users

While individual consumers using GPS navigation devices or mobile apps make up the largest share of the market, commercial users collectively spend more on GPS equipment than the military does. GPS equipment sales across North America averaged $33.5 billion annually from 2005 to 2010.[135] Consumers made 59 percent of those purchases and the commercial segment accounted for 25 percent, while the military represented 16 percent.[136]

Since the mid-1980s, when surveyors began pinpointing terrestrial features with newfound precision, nonmilitary uses of GPS have so proliferated that it is difficult to catalog them, and new applications appear regularly. A recent market forecast offered the following breakdown by sector: aviation, maritime and waterways, highway and construction, public transportation, railroads, communications, emergency response, surveying, weather, scientific, space, environmental protection, recreation and sports, law enforcement and legal services, and agriculture and forestry.[137] This list does not mention the retail, marketing, and advertising industries, but as many commercial applications migrate from expensive stand-alone equipment to smartphone-enabled apps, sellers and ads are sure to follow.

The transition from mapping military targets to using precise coordinates to keep track of municipal infrastructure like utility poles and fire hydrants was a natural one. Civil engineering users were at the forefront of systems employing augmentation techniques such as differential GPS, signals from Russia's GLONASS satellites, carrier-phase signal processors, and lasers to achieve accuracies measured in centimeters. Today companies like Mobile 311, based in Cary, North Carolina, offer smartphone solutions that allow city workers in the field to upload infrastructure coordinates and the locations of sanitation problems, code violations, or repair needs directly to digital geographic information systems.[138] The City of Boston has extended this capability to its residents by creating free apps. Citizens Connect allows users to report and upload photos of graffiti, broken streetlights, and damaged signs.[139] Users can track the progress of the city's response online and view all reports on a map. Street Bump is a crowd-sourcing approach to transportation planning that uses the accelerometers and GPS in drivers' smartphones to detect and upload pothole locations and rough pavement in real time as users travel Boston's streets.[140] Communities of virtually every size today use GIS mapping, and apps like these will undoubtedly become more widespread. U.S. Census Bureau workers used GPS-equipped handheld computers in 2010 to ensure that census reports accurately link data to the proper geographic area. Some residents evidently saw listing their latitude and longitude coordinates as more intrusive than listing a street address. The agency responded to privacy concerns on its website, explaining that by law it may not share information on individual respondents, including GPS coordinates, with any other agency.[141]

GPS accuracy yields better maps, but correcting old problems can create new ones. The North Carolina–South Carolina Joint Boundary Commission authorized new surveys to correct errors dating back to 1735, when surveyors gauged direction from the sun and stars and marked the boundary with wooden stakes and hatchet marks on trees. Near Charlotte the proposed new state line, 150 feet south of the old one, would convert ninety-three South Carolina property owners into North Carolina residents. Convenience store owners who would be affected by the change say it would put them out of business because of North Carolina's higher gas tax and ban on selling fireworks.[142]

Small improvements in precision multiplied over large distances, areas, or volumes in a commercial enterprise can mean huge differences to the bottom line. GPS-enabled machine control has revolutionized such large-scale enterprises as agriculture, construction, dredging, excavation, grading, and paving. In 1996 about 5 percent of farmers used GPS precision to adapt cultivation techniques to soil variability.[143] Instead of spreading seeds, irrigation, fertilizers, and herbicides uniformly across a field, they began managing these inputs down to the square yard and eventually to fractions of an inch using row guidance to avoid costly overlapping of applications. A farmer with 1,600 acres could purchase a GPS farming package from Rockwell or John Deere in 1996 for between $6,500 and $8,900 and save $16,000 on phosphorus and potash alone in the first year.[144] By 2010 about 60 percent of farmers had adopted GPS systems.[145] The average unit price of GPS agricultural equipment was around $13,000, but farmers could cultivate more land with fewer tractors, operating them around the clock in critical planting and harvesting months.[146] Analysts estimate that GPS reduces U.S. farmers' input costs by $9.8 billion annually while generating improved yields worth $10.1 billion.[147] This $20 billion boost to the U.S. agricultural industry amounts to roughly 12 percent of annual production.[148] Some authorities believe precision farming will reach a 100 percent adoption rate within five to ten years.[149]

Trimble Navigation introduced SiteVision GPS Grade Control in 1999, which placed site plans on computer screens in bulldozer cabs, enabling operators to excavate without surveyed stakes in the ground.[150] The technology, developed in open-pit mining, lowered costs, saved time, and improved accuracy in large road construction and land development projects. Precision three-dimensional machine control today shaves waste and boosts profits in a variety of settings where crews previously had to survey, set stakes, and run string lines to maintain proper elevations. At $50 per ton for asphalt, a contractor paving a ten-mile road seventy-two feet wide can save $140,000 by pouring one-one-hundredth of a foot less asphalt.[151] A landfill that charges a tipping fee of $40 per ton can generate an extra $2 million of revenue and extend the site's life span by improving compaction 10 percent across a half-million cubic feet of available airspace.[152] Contractors dredging 130,000 cubic yards of mud from the Ohio River near Mount Vernon, Indiana, in 2011 used a three-dimensional GPS system to "see" their bucket digging on the river bottom, speeding the work and avoiding overexcavation.[153] Analysts estimate that about 40 percent of U.S. heavy and civil engineering construction firms have adopted GPS, producing $9.2 billion in cost savings annually.[154] GPS-guided robots and "smart machines" will increasingly affect other spheres of activity in the future, a topic explored in the next chapter.

Scientists and engineers are adopting GPS for a variety of environmental uses, including monitoring bridges, earthquake fault lines, bird migration patterns, and endangered species. Researchers in the United Kingdom demonstrated in 2006 that GPS sensors could detect bridge movements of less than a centimeter. By establishing a baseline for bridge deflections due to traffic loads, temperature, winds, and deterioration, GPS sensors can alert authorities when bridge movements exceed design tolerances.[155] Russian engineers in 2011 installed sensors with dual GLONASS/GPS receivers on a new 3,887-foot cable suspension bridge—the world's longest—linking the far eastern city of Vladivostok to Russky Island.[156] Such dual-signal receivers roughly double the number of potential satellites visible, making them ideal for locations where tall buildings, tree canopies, or bridge structures mask parts of the sky.

Like surveyors, geophysicists did not need the full complement of satellites to begin monitoring fault lines using GPS receivers bolted to solid rock at surveyed positions in earthquake-prone areas. China installed its first GPS earthquake monitors in 1985.[157] The Southern California Earthquake Center had installed 350 monitors in the region by 1991.[158] A year after the January 17, 1995, earthquake in Kobe, Japan, that killed more than six thousand people, the government doubled the number of GPS monitoring sites across the island nation to about one thousand.[159] GPS sensor measurements before and after the magnitude 8.8 earthquake that struck southern Chile on February 27, 2010, showed that it moved the entire city of Concepción about ten feet west and shifted Buenos Aires, Argentina—eight hundred miles from the epicenter—about an inch.[160] Seismologists announced plans to add fifty sensors to the twenty-five already in the region.[161] The March 11, 2011, magnitude 9.0 earthquake off the coast of Japan that caused devastating tsunami waves intensified the search for faster analysis and earlier warnings. Initial reports underestimated the magnitude as 7.1 because severe shaking often saturates seismometers near large quakes. By the time geophysicists calculated the actual intensity more than twenty minutes later, authorities had lost valuable time needed to alert people ahead of the waves.[162] NASA announced in April 2012 its collaboration with several universities and scientific organizations to test a new warning system, the Real-Time Earthquake Analysis for Disaster (READI) Mitigation Network.[163] It uses satellites to monitor about five hundred GPS ground sensors across California, Oregon, and Washington.[164] Scientists involved in the project say the system could have established the magnitude of the 2011 Japan earthquake within two to three minutes.[165]

Along with precise positioning and navigation, the atomic clock accuracy of GPS time has become an unseen but ubiquitous background utility. More than five hundred thousand timing GPS receivers support industry around the world.[166] They help millions of individuals and businesses to simultaneously download and upload data across private networks and the Internet. Packets of digitized information speed through electronic networks like vehicles on crowded freeways, and precise time synchronization is a prerequisite to maintain smooth flow. Consider the importance of precise time stamping on large electronic financial transactions when market prices are fluctuating rapidly. A clock error of ten parts per million will grow to almost a full second in one day.[167] Network technicians install time synchronizing equipment like Spectracom's SecureSync system or Symmetricom's SyncServer SGC-1500 Smart Grid Clock, which keeps an electric utility's substation clocks synced to GPS time within one microsecond.[168] Precise time synchronization helps power companies manage transmission networks, isolate faults, and balance loads. It is a key element in making our electric grids "smarter" and more secure as utilities begin to integrate dispersed renewable power sources, such as wind turbines and solar cells, with traditional sources, such as centralized coal, gas, and nuclear plants. In the telecommunications industry, most of the world's half-million cellular base stations utilize GPS timing equipment to switch calls (usually seamlessly) from one tower to the next as mobile phone users drive along roadways.[169]

GPS technology is seeping into other commercial and professional activities as users find ways to improve accuracy, increase productivity, lower costs, save time, or enhance safety. While commercial users account for a quarter of GPS equipment sales—less than half the consumer market—the $8.3 billion they spend and the uses they make of GPS yield a large economic impact. Estimates of the aggregate economic benefits to all commercial GPS users in the U.S.

economy range from $68 billion to $122 billion annually, representing between 0.5 and 0.9 percent of gross domestic product.[170]

Delays and Distortions

Consumers, commercial users, and the military have shared a desire for greater precision and helped to drive improvements in GPS technology. From the golfer using a range finder to estimate the distance to a green, to the city worker mapping fire hydrant coordinates in snow-heavy locales, to search-and-rescue crews locating a downed pilot, accuracy is key. One reason the Air Force felt compelled to comment publicly in late 2009 on reports about drivers getting lost using GPS was as part of its ongoing pushback against widespread perceptions that aging satellites were making the system inaccurate. This public relations problem began in May 2009, when the Government Accountability Office briefed the House Subcommittee on National Security and Foreign Affairs about problems in the GPS IIF satellite replenishment program. The IIF generation consists of twelve Boeing-built satellites designed to replace a group of the original operational satellites, some launched between 1990 and 1993 and now far beyond their expected seven-and-a-half-year life spans. The GAO warned that cost overruns and production delays threatened the Defense Department's ability to replace the aging satellites quickly enough to maintain its performance guarantee of twenty-four operational satellites 95 percent of the time. The IIF program was then $870 million over its original $729 million cost estimate, and the first launch, scheduled for November 2009, was about three years late.[171] While commending the Air Force for taking steps to avoid similar problems in the forthcoming GPS III satellite program (which included switching the prime contractor from Boeing to Lockheed Martin), GAO said there was a "high risk" that the GPS constellation would fall below twenty-four satellites between 2010 and 2014.[172]

The hearing itself drew scant media coverage. Major news outlets seemed more focused on the Obama administration's proposal to defund Loran C, a long-range, ground-based radio navigation system that some thought GPS had rendered obsolete.[173] *Government Computer News* posted one of the few stories to appear the next day under the headline "DOD Faces Tough Hurdles in Maintaining, Upgrading an Aging GPS."[174] As the story spread to other media, online and overseas, the headlines grew more dramatic. The science and technology website Gizmodo.com announced, "GPS Accuracy Might Be Less Accurate in 2010," while Engadget.com offered, "GPS System Might Begin to Fail in 2010, Government Accounting Office Warns."[175] The UK-based *Guardian Unlimited* upped the ante with "GPS System 'Close to Breakdown,'" and the newspaper *mX*, in Melbourne, Australia, served up "Satnav Crisis: So Where the Hell Are We?"[176] *RusData DiaLine*, a unit of Russia's *Izvestia*, penned the optimistic headline "Russia's GLONASS to Profit from GPS Problems."[177] Such headlines called to mind the breathless reporting of the Vanguard-Sputnik era and the atmospherics that often accompany stories about technology managed by the military.

The Air Force countered with a public relations offensive, including news releases, interviews, the Air Force Space Command's first use of Twitter for a two-way "tweet forum," and an informational video on YouTube.[178] Col. Dave Buckman, chief of Space Command's Positioning, Navigation and Timing Division, tweeted, "GPS will not go down."[179] In the video Lt. Gen. John E. Hyten, vice commander of Air Force Space Command, explained that thirty-four satellites were in orbit, two launches were scheduled in the coming year, and in the worst year of GPS's history (2000) no more than three older satellites stopped working, providing an ample cushion above the twenty-four-satellite threshold. "It's the healthiest constellation that we've ever had," Hyten noted, "and we are very confident that we'll be able to maintain that level of performance in GPS now and well into the next decade."[180]

Critics noted that despite the large number of orbiting satellites, each with built-in redundancies such as multiple atomic clocks, many older ones were down to "single thread" operation, meaning one more component failure could leave a satellite unusable.[181] Widely reported signal problems with a satellite launched in March, just two months before the GAO uproar, made the Air Force's public relations task more difficult. The satellite, SVN 49, was the last of eight IIR(M) models, which introduced a second civilian signal (L2C). Engineers added equipment to SVN 49 to demonstrate and test the new L5 signal intended for commercial aviation use that would be a standard feature of the IIF series. That change inadvertently affected the other signals, causing a permanent multipath (reflected) signal—a thirty-nanosecond echo—within the satellite itself.[182] The effects of the signal distortion varied with the angle of the satellite in the sky and the type of receiver, complicating potential remedies for the problem.[183] At this writing SVN 49 remains listed as "unusable."[184]

Such problems tended to attract more publicity than a significant advance that occurred about the same time. The Naval Research Laboratory and a team comprising mobile satellite service provider Iridium Satellite, Boeing, Rockwell Collins, Coherent Navigation, and academic experts in 2009 demonstrated the feasibility of the High Integrity GPS program, an augmentation system sometimes called iGPS or HIGPS. The program uses Iridium's constellation of sixty-six low-earth-orbiting satellites to broadcast powerful signals with embedded GPS data, allowing military receivers to lock onto GPS signals faster and maintain them under substantial enemy jamming while in a moving vehicle.[185] Boeing completed the first phase of implementation in 2011, modifying the computers and software that control Iridium satellites, and is now working on the NRL's HIGPS operations center, reference stations, and user equipment.[186]

The Air Force finally launched the first IIF satellite, SVN 62, in May 2010. GPS officials acknowledged in September that during the standard three-month test period they had discovered a software problem in the cross-links that allow satellites to communicate with each other.[187] The problem degraded the system that detects nuclear detonations but did not affect the satellite's navigation and timing signals, so it joined the operational fleet. However, the issue postponed the second IIF launch until 2011, when the software fix would be transmitted to the orbiting satellite.[188] The cross-link software glitch came on the heels of news that surfaced in June about another IIF-related software problem that January. Compatibility issues left as many as ten thousand military receivers useless for several days after crews installed new ground control software for the IIFs.[189] The Air Force delayed the June launch of the second IIF satellite for two weeks and shut off the orbiting satellite's military signal while fixing another issue, described first as an anomaly and later as an electrical problem.[190] Depending on their point of view, observers have blamed the IIF program's troubles and bad publicity on complacency, poor oversight, inherent design flaws, overly ambitious production plans, the challenges of increasingly complex technology, heightened media scrutiny, and gremlins. After all, how many people have set up a new computer without any glitches? But the $121 million spent on average per satellite makes public interest and concern understandable.

Although fears of a complete GPS failure receded, the GAO warning focused broader attention on brownouts.[191] These are temporary periods of an hour or two, widely scattered geographically and throughout the year, when too few satellites are visible for high-precision users such as farmers and heavy equipment operators. Having thirty satellites in the constellation for several years had reduced this inconvenience. If the constellation drops to twenty-four or fewer satellites, users would more often notice a difference, particularly when trees or buildings block one or two satellites. Many high-end GPS receivers have dual-frequency capability, like the bridge sensors mentioned earlier, to take advantage of any additional Russian GLONASS satellites in view. The brownout issue illustrates how users benefit from other satellite navigation systems now joining GPS in orbit. However, other GPS-like systems—often referenced generically as GNSS, for any *global* navigation satellite system—spark competition and occasional conflict.

Everybody Wants One

As mentioned in the preceding chapter, GLONASS began in 1982 under the Soviet regime and attained twenty-four operational satellites in 1995. It declined precipitously in the latter half of the 1990s because each satellite had only a three-year design life.[192] GLONASS uses a different type of radio signal but otherwise is very similar to GPS. Some have asserted that the Bank of Credit and Commerce International (BCCI), which collapsed in 1991 amid a scandal involving financial fraud, money laundering, illicit technology transfers, arms sales, and drug trafficking, sold Soviet spies secret NAVSTAR GPS plans.[193] A source in the investigation showed *Time* magazine reporters various photographs he claimed to have made of NAVSTAR technical documents in the Kremlin, but it is unclear when or how the Soviets acquired them.[194] While the U.S. military's signal encryption codes are top secret, some aspects of GPS have been public and published for commercial users from the start. Russian Federation president Boris Yeltsin opened GLONASS for civilian use in 1999, but the *Moscow Times* described the fleet as being "in a dismal state"—nine satellites in orbit and only six working—when the Russian Cabinet in August 2001 approved spending 23.6 billion rubles (roughly $800 million at the time) to rejuvenate the constellation over ten years.[195] The second-generation satellite, designated GLONASS M, added another civilian frequency, improved the antennas, and more than doubled the life span, to seven years.[196] Russia in 2007 announced GLONASS K, a smaller, lighter version with a third civilian signal and twelve-year design life, offering the option to launch as many as six on a single rocket.[197] In December 2011 the GLONASS constellation again reached twenty-four operational satellites, and at a March 2012 navigation conference in Munich, Germany, a Russian Space Agency official confirmed his nation's desire to expand the constellation to thirty.[198] The Russian government has approved 347 billion rubles (nearly $12 billion) through 2020 for new satellites, a ground-based differential augmentation system, and a new operational control system.[199] A 2006 agreement gave India access to GLONASS military signals and gave the Russian program a cash infusion.[200] Russia offered India full joint partner status in mid-2012.[201] Russian efforts at international cooperation have included making GLONASS signals compatible with GPS signals and committing to have every GLONASS satellite by 2020 broadcast code division multiple access (CDMA) signals, the format GPS and other systems use (see chapter 5), along with its native format, frequency division code modulation (FDMA). However, Russia sometimes plays hardball. In a move presumably backed by President Vladimir Putin, Moscow threatened in mid-2012 to block imports of mobile phones and other devices that do not have dual GPS/GLONASS chipsets.[202]

The European Union's Galileo system is about five years behind schedule. Proposals for a civilian-run European alternative to GPS predate by a decade the 1993 Maastricht Treaty establishing the EU. The European Space Agency (ESA), with eleven members in 1983, proposed a constellation of twenty-four satellites at a time when budget woes had U.S. officials scaling GPS back to eighteen.[203] ESA officials thought they could build their system for about $2.5 billion, half the GPS estimate, and begin launching satellites in 1988, the year U.S. officials anticipated completing GPS.[204] By the early 1990s European plans envisioned a public-private partnership building a thirty-satellite constellation offering a mix of free basic service and fee-based precision landing guidance.[205] After EU leaders in Brussels, Belgium, approved the first funds in 1999 for a $3.06 billion program cosponsored with ESA, it took five years to reach an agreement with U.S. military officials, who were concerned about Galileo's signals interfering with GPS military signals.[206] China invested €200 million (about $270 million) in Galileo in 2003 but complained that it felt shut out of decision making after private sector funding collapsed. The EU rescued the program with public funds, which tended to elevate political and security concerns.[207]

China launched its first Beidou (sometimes called Compass) satellite in 2007, apparently seeking to leverage negotiations with the EU, but talks failed.[208] In April 2009 China launched its second Beidou satellite, effectively ending its relationship with the EU, and announced that the Beidou system would use some radio frequencies Galileo had reserved for encrypted governmental, public safety, and possible military use.[209] It was an aggressive but legal move. The United Nations International Telecommunications Union, which allocates radio spectrum for satellites, grants priority to the first country establishing services at a specific frequency.[210] It is unclear whether the EU could have salvaged the partnership. Later that year a retired People's Liberation Army colonel told a reporter that since the 1996 missile crisis in the Taiwan Strait, China had been committed to building its own GNSS. During the standoff the Chinese army fired three GPS-guided missiles toward Taiwan as a warning against seeking independence, and the second two missiles went awry. Military officials suspected someone disrupted the GPS signals, and the retired colonel called the incident an "unforgettable humiliation."[211]

China launched its sixteenth Beidou satellite in October 2012 and began offering service across the Asia-Pacific region in late December 2012.[212] China plans to have thirty to thirty-five satellites in orbit by 2020, providing worldwide service.[213] Beijing announced in August 2012 that it would build eight regional testing centers within three years to certify civilian equipment and five development hubs to spur innovation, incubate new enterprises, and accelerate industrial adoption.[214] There are an estimated 30 million GPS navigation devices in use in China, compared to about 120,000 civilian and military Beidou users so far.[215] Chinese officials say they will gain market share with lower prices and better accuracy.[216]

Europe's ESA meanwhile launched Galileo prototypes in 2005 and 2008, the first two operational satellites in 2011, and two more in late 2012.[217] A fact sheet available at the ESA website provides an overview of the program: The agency currently launches satellites in pairs aboard Russian-made Soyuz rockets but plans to begin launching four at a time aboard Arianespace rockets starting in 2014. All launches take place from the European spaceport in French Guiana. After the first four satellites complete in-orbit validation tests, the schedule foresees eighteen satellites in orbit by 2015 and thirty by 2020. The Galileo constellation will have twenty-seven satellites and three spares in three orbital planes angled fifty-six degrees above the equator. They orbit a bit higher than GPS satellites, taking fourteen hours to circle the planet.[218] Each satellite carries two atomic clocks—a rubidium frequency standard and a hydrogen maser, designated as the master clock.[219] Galileo will offer five services. A basic open service, like GPS, will be free of charge for all users. Subscriber services include an encrypted commercial service with higher data throughput and better accuracy, a safety-of-life service with a built-in integrity feature to alert aviation and maritime users of system failures, and a public regulated service with controlled access for such users as police and first responders. It will also provide search-and-rescue service linked to an international distress system.[220]

Each Galileo satellite will broadcast ten coded signals.[221] Galileo's Open

Service and GPS are supposed to share a new common civilian signal, E1/L1C, on the same frequency by using a technique called multiplex binary offset carrier (MBOC), developed by a joint task force during the negotiations that produced the 2004 U.S.-European agreement. However, a transatlantic tempest erupted in April 2012 after the United Kingdom awarded a patent for MBOC to a wholly owned subsidiary of the British military's UK Defence Science and Technology Laboratory.[222] It listed the inventors as two British engineers who were part of the joint task force.[223] A patent would force manufacturers to pay royalties and raise the prices of receivers designed for a taxpayer-funded free service. Even the U.S. military would pay more because it purchases many civilian devices for noncombat missions.[224] The patent dispute appeared headed for resolution after the British secretary of state for defence in October 2012 asked the European Patent Office to revoke the patent.[225]

There is also the matter of Galileo's funding. Cost overruns, together with slow economic growth, monetary problems, and austerity measures among member nations, remain a challenge. The European Parliament approved €3.4 billion in late 2007 to cover the entire program through 2013, shifting a substantial sum from agriculture in the process.[226] The total covers the first fourteen satellites and the ground control segment as well as the operation of the European Geostationary Overlay System (EGNOS), Europe's version of the U.S. WAAS satellite augmentation system. Officials estimate that combined operating costs for Galileo, after completion, and EGNOS, declared operational for aviation in 2011, will be about €800 million annually.[227] EU lawmakers began drafting proposals in spring 2012 allocating €7 billion (nearly $9 billion) for the program over the next EU budget period, 2014-20, to reach full operational capability with thirty satellites.[228] In February 2013 the European Council trimmed the commission's proposal by 10 percent, approving €6.3 billion for Galileo.[229] However, the European Parliament, the third body in the EU's complicated budget-making system, had not weighed in on the figure as this book went to production. The European Commission, which proposes and administers EU policies, estimated that satellite navigation accounted for about 7 percent of EU gross domestic product in 2009.[230] It cited independent studies projecting €90 billion in direct and indirect economic benefits to the EU over Galileo's first twenty years of operation.[231] As part of the commission's efforts to boost public enthusiasm, it sponsored an art competition for schoolchildren ages nine to eleven, with a satellite being named after the child whose drawing placed first in each nation.[232]

Other navigation satellite systems being developed are regional in scope. The Japanese government in 2002 authorized the Quasi-Zenith Satellite System (QZSS), led initially by a group of businesses that later pulled out of the project and were replaced in 2007 by the government-sponsored Japanese Aerospace Exploration Agency.[233] It launched one satellite in 2010, plans three more launches by the end of the decade, and envisions a constellation of seven satellites eventually.[234] The government accelerated development in fall 2011, apparently motivated by the earthquake and tsunami.[235] The system's name derives from its highly elliptical orbits, which will keep three satellites over Japan more than twelve hours a day and at least one almost directly overhead at most times.[236] This placement facilitates better visibility for car navigation in the urban canyons of Japanese cities.[237] Japan also operates the Multifunctional Transport Satellite Augmentation System (MSAS), which, like WAAS and EGNOS, uses satellites to transmit signals enhanced by a network of ground reference stations.[238] It became operational in 2007.[239]

Less than a week after Russia offered India a joint development stake in upgrading GLONASS, India announced that it would launch in 2013 the first of seven satellites for its Indian Regional Navigation Satellite System (IRNSS).[240] Although the two nations entered discussions at the end of 2010 about closer ties between their programs, it is unclear how India will proceed.[241] Indian media have tended to highlight ending dependence on GPS and to overlook IRNSS's regional limitations with such headlines as "Scientists Excited about India's Own GPS."[242] The government began the program in 2006, and the Indian Space Research Organization expects to complete the constellation in 2014, although the original first launch was supposed to occur in 2012.[243] IRNSS will have a different architecture from any other GNSS. Three satellites will have geostationary orbits, traveling at the same rate as the earth's rotation, so they will appear fixed in the sky like those used for satellite television. The other four will have geosynchronous orbits, circling the earth in twenty-four hours, but because they will be inclined twenty-nine degrees from the equator, their movement in the sky will resemble a figure eight.[244] This arrangement will keep all seven satellites visible in the Indian region twenty-four hours a day.[245] India, too, is developing a space-based augmentation system. The GPS-Aided Geo-Augmented Navigation system, or GAGAN, will consist of three geostationary satellites linked to a network of ground stations, which are in their final testing phase.[246] After an initial launch failed in 2010, India placed the first GAGAN satellite in orbit in May 2011 and the second in September 2012.[247]

Not to be outdone, Iran announced plans in May 2012 to develop Nasir 1, described as a "domestically designed and manufactured satellite navigation system."[248] The director of Iran's space agency said Iran was among a handful of nations capable of developing satellite technology but provided no details on the number of satellites, system design, or launch dates beyond hopes for "the near future." Curiously, the second sentence of the widely circulated report by the Mehr News Agency stated, "The satellite navigation system has been designed to find the precise locations of satellites moving in orbit." It is unclear whether this is a translation issue, incomplete reporting, or veiled bravado. Iran two weeks earlier had scheduled the launch of its Fajr (Dawn) reconnaissance satellite to coincide with the start of nuclear negotiations with six major powers.[249] That day passed without a launch, and Iran announced a one-year postponement the day before the unveiling of its satellite navigation aspirations.[250]

U.S. leaders can draw many lessons from what other nations or groups of nations are doing with GNSS. Here are three: First, delays caused by technical and financial challenges are common. Second, the United States cannot be complacent about upgrading GPS, because other GNSS developers are numerous, determined, and capable. Third, the United States must compete and cooperate simultaneously—practicing "co-opetition"—and the best place to do that is from the lead.[251] Thus, government officials frequently discuss maintaining GPS as the global "gold standard."[252]

At this writing four GPS IIF satellites are in orbit and the prime contractor, Boeing, projects delivery of the twelfth and final IIF satellite in 2013.[253] The Air Force plans to launch two IIF satellites per year through 2016.[254] The IIF's new features, such as the L5 signal, which the FAA will use for new flight procedures that accommodate more traffic and save fuel, require a sufficient number of satellites in orbit. The same situation existed before the original GPS constellation was completed, but with forthcoming Galileo satellites offering compatible L5 signals, there could be as many as thirty combined satellites broadcasting L5 by the end of 2015, rather than by 2019 under the GPS schedule alone.[255]

For several years, the Air Force insisted that the first GPS III satellite would launch in April 2014. (The next chapter discusses GPS III features in detail.) Lockheed Martin officials have said their company is on track to deliver the first GPS III satellite "flight ready" in 2014.[256] The IIF delays that GAO criticized created this significant overlap of launch schedules. Fortunately, the Air Force has been able to manage the aging constellation in a manner that makes it unlikely that the number of satellites could fall below twenty-four before 2014.

Having a backlog of satellites ready to go is obviously better than needing satellites that are not ready, but scheduling payloads on launch vehicles in available time slots could become an issue. That possibility is lessened because the Air Force is designing the block IIIs to fly aboard either Delta IV or Atlas V rockets and to be launched in pairs, either together or with other "dual manifest" satellites.[257]

More importantly, although the first GPS III satellite will be available for launch in April 2014, a new ground control system designed specifically for it will not be ready. Replacing the existing Boeing-built Operational Control System (OCS) with a Raytheon-built next generation Operational Control Segment (OCX) is about seventeen months behind schedule, and some independent observers estimate that upgrades required to fully utilize GPS III's advanced features will not be complete until mid- to late 2017.[258]

OCX is being phased in. The initial phase will provide launch and checkout capability for GPS III satellites. The second phase will merge command and control of older satellites and GPS IIIs, including new civilian signals. The third phase will accommodate new international and modernized military signals. The OCX delay has compelled the Air Force to postpone the first GPS III launch until May 2015, a schedule that depends on no additional problems slowing OCX development.[259]

Homegrown Static

Amid the IIF delays, brownout worries, and out-of-sync space and ground segment upgrades, the GPS industry suddenly faced a new challenge. A company named LightSquared announced plans to build a nationwide 4G mobile broadband network using radio spectrum adjacent to the frequencies GPS uses. LightSquared formed in early 2010, when a New York hedge fund, Harbinger Capital Partners, acquired mobile satellite service provider SkyTerra.[260] The FCC had given SkyTerra (under its previous name, Mobile Satellite Ventures) permission to build an ancillary terrestrial component—cellular ground stations—to augment coverage where satellite signals do not work well, such as urban areas. That authorization allowed a maximum of 2,415 ground stations in the United States.[261] LightSquared conceived a plan to develop a network of forty thousand ground stations and provide wholesale broadband to wireless carriers—making almost the entire nation a Wi-Fi hotspot. It signed an eight-year, $7 billion deal with Nokia Siemens Networks in July 2010 to build the network and asked the FCC for approval in November.[262]

What had been a mobile satellite service with a limited number of ground stations posing scattered risks of GPS interference morphed into a predominantly ground-based network with a minor satellite component posing potential interference problems nationwide. GPS experts said that LightSquared's ground station signals, by some estimates five billion times more powerful than faint GPS signals from space, would disrupt receivers on the ground nearly six miles away and receivers in aircraft up to twelve miles away.[263]

In late January 2011 the FCC gave LightSquared permission to proceed on the condition it allay interference concerns expressed by the Commerce Department's National Telecommunications and Information Administration (NTIA) and the U.S. GPS Industry Council.[264] "Seething" is how *Inside GNSS* reporter Dee Ann Divis described the reactions of people in the GPS community.[265] When the FCC ordered LightSquared and industry representatives to form a joint working group, conduct interference tests, and present findings within five months, many perceived the FCC to be rushing the process, driven by political cronyism.[266] The Center for Public Integrity, nonpartisan but sometimes accused of leaning left, compiled a report calling the connections between LightSquared and the Obama administration a clear example of political patronage.[267] Among their findings: Billionaire Philip Falcone, head of Harbinger Capital, had previously contributed the maximum allowable to both Democratic and Republican Senate campaign committees, ensuring access. FCC chairman Julius Genachowski was one of the Obama 2008 campaign's major fundraising "bundlers" and served on the Obama transition team with Jeff Carlisle, LightSquared's executive vice president for regulatory affairs, creating a too-cozy relationship between the regulator and the petitioner seeking a waiver.

As technical studies proceeded in laboratories, the controversy played out in a public battle that lasted more than a year and attracted broader media coverage than perhaps any GPS issue to date. Both sides hired lobbyists and public relations firms.[268] A group of industry leaders formed the Coalition to Save Our GPS in March 2011, gathering support from aviation, agriculture, transportation, construction, engineering, surveying, GPS equipment manufacturers, and service providers.[269] Their efforts along with the concerns of federal agencies and the military soon brought elected officials into the fray. Thirty-four senators asked the FCC to rescind the waiver, and the House passed the National Defense Authorization Act (NDAA) with a provision prohibiting the FCC from giving LightSquared final approval until the Defense Department's concerns were satisfied.[270] President Obama signed the NDAA and later included similar language in his budget concerning interference of commercial devices.[271]

Responding to criticism from such major companies as General Motors and John Deere., LightSquared's public relations campaign ran nationwide full-page ads in the *New York Times* and *Wall Street Journal* and launched the supposedly grassroots Empower Rural America Initiative to bring broadband to the heartland.[272] The tone of the company's news releases ranged from cooperation to contempt. After tests confirmed interference, LightSquared offered in June 2011 to swap spectrum with another Harbinger-affiliated satellite provider and use a frequency farther from GPS signals at lower power until a technical solution emerged.[273] When GPS manufacturers balked because they were unsure any technical remedy was possible, LightSquared blamed the industry for failing to build receivers with adequate noise filters and called the billions spent by government on GPS a "federal handout" to them.[274] The FCC ordered more tests.[275] In October the company's general counsel released a statement that seemed to lay the groundwork for litigation, and Carlisle, the executive vice president, accused National PNT Advisory Board members of bias due to their financial interests.[276] As the January 26 anniversary of the FCC waiver approached, the company charged that GPS industry insiders had "rigged" test results by using obsolete receivers.[277] Intense lobbying continued. The NTIA sent the FCC a letter on February 14, 2012, summarizing its review of tests by various federal agencies. "We conclude that LightSquared's proposed mobile broadband network will impact GPS services and that there is no practical way to mitigate the potential interference at this time," it stated.[278] Later that day the FCC issued a news release announcing its intention to vacate the conditional waiver and suspend indefinitely LightSquared's authority to build more ground stations.[279] Responding to the decision, LightSquared CEO Sanjiv Ahuja characterized the nationwide coverage requirements the FCC attached to the company's original request as a government mandate that forced LightSquared to invest nearly $4 billion only to have the agency change its mind.[280] The company vowed to fight on, but its finances were unraveling. Within two weeks a group of investors sued Harbinger Capital, and LightSquared defaulted on a $56 million payment to a satellite operator.[281] A week later the company cut nearly half of its 330 employees and Ahuja stepped down as CEO.[282] In March Sprint Nextel canceled a $9 billion, fifteen-year spectrum hosting agreement, and on May 14, 2012, LightSquared filed for bankruptcy, still vowing to reorganize and find a way to build its network.[283]

The LightSquared controversy was costly and disruptive but may have long-term benefits. It significantly raised public awareness of the role GPS now plays in the U.S. economy and the nation's security and prompted the National PNT Advisory Board to develop a series of recommendations about jamming. They include urging the executive branch to designate GPS as critical national infrastructure subject to Department of Homeland Security oversight, creating a system for reporting locations with persistent interference and eliminating them, cracking down on GPS jammers, strengthening receivers and antennas, and funding a national backup capability for PNT needs.[284]

Whereas previous economic studies of GPS focused on market opportunities and sales forecasts, the Coalition to Save Our GPS commissioned the first analysis of what a GPS disruption might cost the economy. NDP Consulting Group, based in Washington DC, submitted its report *The Economic Benefits of Commercial GPS Use in the U.S. and the Costs of Potential Disruption* in June 2011. (Many figures cited in this book come from that study.) NDP focused on three commercial sectors with high GPS adoption rates and readily available data—precision agriculture, engineering/construction, and surface transportation—to project the impact on commercial users. The study found that full nationwide disruption of GPS would produce a direct economic impact of $96 billion per year—about 0.7 percent of the U.S. economy.[285] Given that the commercial market is only a quarter of the total GPS market, that shows the significant return from the roughly $35 billion in public funds spent on the constellation to date and the nearly $1-billion-per-year cost of maintaining GPS.[286]

With increasing competition for limited radio spectrum, the National PNT Advisory Board expects more interference threats and wants to be better prepared to defend GPS. At its August 2012 meeting, members discussed commissioning a more comprehensive study of the economic benefits that GPS produces for the United States and the rest of the world.[287] That promises to be an enormous number-crunching task, given the scope of GPS use today, and one made harder by the prospect of accounting for the growing use of receivers that combine two, three, or four different GNSS signals. The European GNSS Agency in a spring 2012 report estimated that the global GNSS market would reach €244 billion, or about $277.4 billion, by 2020. Other reports may arrive at different figures, but they are all likely to be huge numbers.

This chapter began with the question: Where are we? One answer might be, recalculating. In the United States and much of the world, both the culture and the economy are inextricably dependent on accurate, uninterrupted GPS service. Millions of people take it for granted with little or no understanding of its operation or potential problems. Recent events should be a wake-up call. Taken as a whole, the challenges associated with preserving privacy, adding new satellite signals, synchronizing space and ground segment upgrades, avoiding interference by other radio spectrum users, and cooperating with other nations in a GNSS world underscore how complicated maintaining GPS is and how vital a broader public understanding of it has become.

Going Forward
The Future of GPS

If you don't know where you're going, you'll probably end up
somewhere else.

David P. Campbell, book title, 1976

As bestselling author Henry Petroski, a professor of engineering and history
at Duke University, notes in *Pushing the Limits: New Adventures in Engineering,*
"Predicting the technological future has always been risky business, for the
world of invention and engineering never ceases to push the limits of technology to come up with surprises that surprise even the experts."[1] What follows
is less an exercise in prediction than an exploration of what seems certain,
what appears likely, and what *could* happen in the future.

In the near term the variable is "when" rather than "if." The third major
block of satellites, GPS III, should begin launching in 2015. Whereas IIF satellites are replacing the older IIA satellites (eight active as of June 2013), the GPS
III satellites are slated to replace the somewhat newer IIR versions. Even if the
schedule slips, as has so often happened, they will certainly join the constellation within a few years. GPS III offers many technological advances, starting

with the signals. Assuming the United States and Great Britain resolve the patent issues discussed in the preceding chapter, GPS III will introduce a fourth civilian signal (L1C) that is interoperable with Europe's Galileo, Japan's QZSS, and potentially with China's Beidou.[2] Navigational accuracy will be three times better than with the Block IIF satellites. The GPS III's user range error over twenty-four hours will be within one meter, versus the current three meters.[3] GPS III will broadcast military signals three times more powerfully, making jamming by an adversary more difficult.[4] Lockheed Martin's subcontractor on the navigation payload, ITT, has participated in building that component in every generation of GPS satellites for more than thirty years.[5]

How soon the new satellites will deliver these capabilities depends not only on the launch schedule and completion of the new OCX ground control system but also on fielding new receivers designed to take advantage of the new features. GAO in March 2012 warned, "User equipment is not expected to be fully fielded to the warfighters until many years later, possibly as late as 2025."[6]

Lockheed Martin, which built twenty-one Block IIR satellites and "modernized" the last eight of them (IIR[M]), designed GPS III from the start to insert new capabilities as future space vehicles (SVs) are built.[7] Although the Air Force to date has ordered eight GPS III satellites, the 2008 contract has an option for up to a dozen. Lockheed Martin publicity materials indicate the Air Force may build as many as thirty-two GPS III satellites.[8] The ninth and later SVs will support the Distress Alerting Satellite System, a NASA-led international program using GPS and Galileo satellites to relay signals from emergency beacons to search-and-rescue teams.[9] By replacing a current analog component with a digital waveform generator, GPS III satellites will be equipped to generate new types of signals after they are already in orbit.[10] Each GPS III will carry three rubidium atomic frequency standards but will have an open slot for a fourth atomic clock. For the first time, ground controllers will be able to turn on and monitor this backup clock separately from the operational clocks, giving them the ability to study experimental designs, such as a hydrogen maser.[11] These expandable features are important because the design life of each GPS III satellite is fifteen years. The record of technological change during any fifteen-year period since the satellite age arrived suggests how quickly high-tech equipment can become obsolete. If GPS III longevity matches that of earlier generations, some satellites launched in the 2015–16 period could still be orbiting and operational in 2038 or beyond.

GPS III satellites may repeat the pattern of GPS IIA, IIR, IIR(M), and IIF, with

Fig. 10.1. GPS III satellite. (Courtesy National Coordination Office for Position, Navigation and Timing)

enhancements coming in block buys of a dozen or so satellites, or they might pioneer an unprecedented change in the constellation architecture. As this book went to production Air Force officials were considering alternatives they hope could cut costs without stifling technological progress. One proposed scheme would maintain a thirty-plus satellite constellation by combining fifteen to eighteen full-featured satellites with a balance of smaller versions that provide only basic navigation signals. Limiting some satellites to only four signals and forgoing nuclear detonation detection capability would trim weight and power consumption and could save about $200 million per satellite.[12] Although the concept has been floated in the past, the idea may find new support as a way to replenish a mature constellation in the era of budget sequestration.

A key feature of GPS III satellites and the new OCX control system will be the

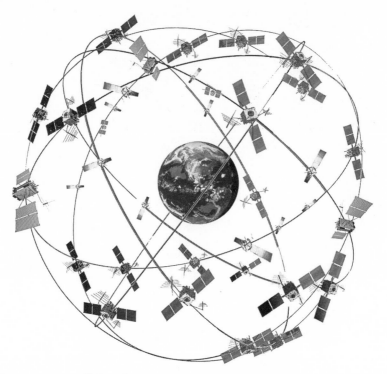

Fig. 10.2. GPS constellation, circa 2015. (Courtesy Lockheed Martin)

ability to operate more than thirty-two active satellites in the constellation. The existing thirty-two satellite limit stems from the number of unique digital identifiers, the PRN codes (described in chapter 5), that were programmed into the original system. There were thirty-seven PRN sequences, with PRNs 1 through 32 assigned as SV ID numbers and 33 through 37 reserved for other uses. Those latter five PRNs are being redesignated for use by SVs, and the new control system will add twenty-six additional codes, PRN numbers 38 through 63.[13] While nobody expects sixty-three GPS satellites in orbit anytime soon, the new technical specifications make that number possible.

The More the Merrier?

GPS started with a constellation of twenty-one satellites and three spares. Over time, users became accustomed to about thirty, and that threshold appears set to grow. The Europeans adopted thirty as the size of the Galileo constellation before the first launch. Russia, having restored GLONASS to twenty-four, wants to expand to thirty. China is working toward a constellation of thirty to thirty-five Beidou satellites. Add to all of these India's IRNSS, Japan's QZSS, and numerous satellite-based augmentation systems (WAAS, EGNOS, et al.), and by 2020 there could be about 150 navigation satellites broadcasting signals in the GPS bandwidth. This redundancy should be an unqualified boon to users, reducing the vulnerability from any single system failure while eliminating brownouts and poor reception in urban canyons, should it not? Unfortunately, the answer is no. Each additional GNSS signal increases the overall background noise that receivers must sort through to lock onto a particular signal. It is like trying to follow a conversation in a crowded room where everyone is talking at once. Guenter Hein, head of ESA's Galileo Operations and Evolution Department, has calculated that once the number of combined signals approaches seventy, the noise from satellites will exceed the natural cosmic noise floor, and GNSS systems will begin to interfere with themselves.[14] While this portends an eventual international squabble over the total number and allocation of GNSS satellites and signals, the primary worry in the near term remains terrestrial interference. This can take the form of inadvertent disruptions by adjacent radio spectrum users like LightSquared or deliberate jamming by pranksters, people trying to evade tracking, and even terrorists.

As noted previously, jamming GPS with inexpensive transmitters is relatively easy because the signals are so faint. Some have compared using a receiver on the ground to detect a GPS signal from space to standing in Los Angeles and trying to see a sixty-watt lightbulb in New York.[15] An example of the serious ramifications of jamming surfaced in 2010, when the FAA investigated why a new ground-based augmentation system (GBAS) at Newark Airport randomly lost GPS signal reception and shut itself down multiple times per day over several months.[16] GBAS systems improve GPS accuracy to allow Category I precision landings, but officials could not approve the Newark system until determining the cause of the outages. The GBAS antenna was within about a few hundred feet of the New Jersey Turnpike, and the cause turned out to be passing truck drivers using inexpensive GPS jammers to thwart their fleet managers' efforts to track their every move.[17] Experts estimated that one hundred thousand such devices, which plug into vehicle cigarette lighters and can cost as little as thirty dollars, were in use in the United States by 2009.[18] Marketing, selling, or using devices that jam signals such as GPS, cell service, police radar, or Wi-Fi is a federal crime in the United States, and the FCC announced a crackdown in October 2011.[19] However, overseas sellers continue to offer a wide array of jammers on the Internet.

North Korea jammed South Korean GPS reception for three days in August 2010, using transmitters that military analysts believe the North Koreans imported from Russia, mounted on vehicles, and stationed near the border.[20] The jamming resumed in May 2011, apparently aimed at disrupting a joint military exercise with the United States.[21] After North Korea's failed attempt to launch a rocket supposedly carrying a satellite in spring 2012, South Korea reported jamming over a two-week period that affected commercial shipping and more than six hundred domestic and international airline flights.[22] The jamming stopped after South Korean president Lee Myung-bak enlisted the help of Chinese president Hu Jintao during a summit meeting.[23] South Korea vowed to raise the issue with the International Telecommunications Union and International Civil Aviation Organization, but North Korea denied it was the source of the interference.[24] The incident illustrates the potential for geo-political rivals to harass their opponents by disrupting satellite signals.

Spoofing potentially poses a bigger threat than jamming since counterfeit GPS signals could go undetected until a serious problem occurred, whereas jamming is apparent due to loss of the signal. The topic grabbed international attention after Iran claimed it used spoofing to bring down a U.S. drone (a stealthy RQ-170 Sentinel unmanned aerial vehicle, or UAV) in December 2011. Iran said its engineers hacked into the drone's electronics and fed it false coordinates, making the UAV think it was landing at an airfield in Afghanistan.[25] U.S. officials disputed the claims, saying a malfunction was more likely. They pointed out that GPS merely backs up the RQ-170's inertial navigation system, correcting drift, and asked why it took the Iranians days to find the downed craft, if they sent it false coordinates.[26] Mechanical and engine failures are common causes of drone crashes, along with lost data links—like a dropped call on a cell phone.[27] Ohio-based air national guardsmen lost control of a $3.8 million Predator drone in September 2012, and Air Force superiors ordered them to crash it into in an Afghan mountain, marking the one hundredth drone lost since 2007.[28] However, GPS specialist Todd Humphreys, an assistant professor of aerospace engineering at the University of Texas at Austin, announced soon after the Iranian incident that he had developed a powerful spoofing device for less than $1,000.[29] He and his students later demonstrated it to Department of Homeland Security officials at White Sands Missile Range in New Mexico, where from about a half mile away, they commandeered a commercial quadrotor helicopter UAV using GPS as its sole navigation system.[30] Humphreys called the Iranian claim "within the realm of possibility" and warned that inexpensive spoofers would be attractive not only to terrorists but to anyone wanting to fool GPS systems—from fishing boats working off-limits waters to financial traders exploiting split-second time discrepancies in fast-moving markets.[31] As GPS adoption increases, spoofing seems likely to prompt new regulations, require new law enforcement techniques, and spawn a black market in the devices.

GPS and GNSS systems have vulnerabilities beyond malevolent jamming and spoofing. The Royal Academy of Engineering in London released a study, *Global Navigation Space Systems: Reliance and Vulnerabilities*, in March 2011 that echoed the U.S. National PNT Advisory Board's call for a backup system for GPS. The report warned that widespread reliance on GNSS signals and the possibility that many seemingly unrelated services could fail simultaneously because of signal disruption have created "an accidental system with a single point of failure."[32] Among potential causes of failure, it listed uploads of bad navigational data, clock anomalies, loss of satellites due to the orbital environment, atmospheric problems, attacks on the ground segment, faulty system upgrades, and receiver bugs. The report cited bad data uploads to GPS satellites in 1993, 2000, and 2002 and clock anomalies in 2001 and 2004. In the 2004 example, the clock error of SVN-23 (the oldest GPS satellite in orbit) went undetected for three hours, producing a user-range error of about 186 miles (300 kilometers) before ground controllers spotted the problem.[33] Fortunately, these incidents caused no serious problems, but they suggest ways that larger failures might occur.

One potential cause of catastrophic failure listed in the report is impossible to predict but has sufficient probability to prompt constant vigilance of the sun. A super solar flare, or Carrington Event, named for English astronomer Richard Carrington, would saturate the orbital environment with highly charged particles that could destroy the electronics aboard satellites. Carrington on September 1, 1859, observed and made sketches of unusually intense sunspots—known as solar flares or coronal mass ejections. The next day telegraph systems worldwide "went haywire," discharging sparks, shocking operators, and setting telegraph paper on fire.[34] Such solar activity went unnoticed before humans created electronic networks. Subsequent solar flares not nearly as large have caused significant damage to telecommunications equipment and power systems. A solar storm disabled the U.S. WAAS network for thirty hours in 2003.[35] X-rays disrupted GPS for about ten minutes in 2005.[36] Researchers have studied chemical evidence of ancient solar storms recorded in arctic ice and the patterns since 1859 to try to predict the frequency of a Carrington-level solar flare, with estimates ranging from once in two hundred to five hundred years.[37]

Sharp Elbows in Space

International conflicts on the ground seem increasingly likely to escalate to space, given today's strategic reliance on navigation, surveillance, and communication satellites. Even if governments avoid confrontations that threaten satellites, avoiding accidents will require better systems and improved cooperation. An unclassified January 2011 summary of the National Security Space Strategy prepared by the Department of Defense and the National Intelligence Agency described the situation this way: "Space is becoming increasingly congested, contested, and competitive."[38] Since the Vanguard satellite launched in 1958 (it is still orbiting), the number of man-made objects in Earth orbit that the Defense Department tracks has increased to about 22,000, including around 1,100 active satellites and about half as many rocket bodies.[39] The military actively monitors objects ten centimeters (about four inches) in diameter and larger because a direct hit could destroy a satellite.[40] However, NASA estimates there are a half-million particles between one centimeter and ten centimeters in diameter in low Earth orbit—below 1,250 miles (2,000 kilometers)—that could damage satellites, the Hubble telescope, or the International Space Station in an impact.[41] Fortunately, GPS satellites are in mid-Earth orbits ten times that distance, but they will be just as vulnerable if present trends creep to higher altitudes.

Debris has increased dramatically over the past few years because of satellite breakups and collisions and an antisatellite (ASAT) test by China. In 2009 an inactive Russian government Cosmos satellite collided with an active U.S. commercial Iridium satellite about 497 miles above Siberia, destroying both and adding 1,500 to 2,000 pieces of space debris.[42] It was the first reported collision between two satellites in orbit, although a rocket fragment struck and damaged a French satellite in 1996.[43] After the 2009 incident the U.S. military established a warning system for close approaches of all satellites it tracks, and the European Union proposed an international code of conduct for outer space.[44] U.S. opponents balked at any agreement that might limit military options, but the Defense and State Departments eventually endorsed the concept in January 2012, expressing confidence that they could create a code of conduct that did not constrain national security–related space activities.[45] That announcement came days after a defunct Russian spacecraft crashed into the Pacific Ocean near Chile.[46]

U.S. officials raised alarms about China's antisatellite activities in 2006, when on at least two reported occasions China aimed a ground-based laser at a U.S. reconnaissance satellite.[47] Sensors on Kwajalein Atoll traced the source of the laser to mainland China.[48] In January 2007 China used a ballistic missile warhead to destroy one of its obsolete weather satellites, creating more than three thousand pieces of new debris.[49] The strike, which required hitting the satellite as it traveled at nearly 17,000 mph, signaled that China's antisatellite technology had advanced beyond that of the former Soviet Union.[50] A year later the United States used a sea-based Navy Aegis ballistic missile interceptor to shoot down a dead spy satellite that was tumbling out of orbit. Officials said it was necessary to ensure that one thousand gallons of toxic thruster propellant did not survive reentry and harm human health.[51] Critics at home and abroad speculated that the United States, which had not conducted any antisatellite exercises since the 1980s, contrived the action to send a message to China.[52] However, unlike the Chinese incident, there were numerous public explanations in advance of the shoot-down, and the impact occurred at an altitude of about one hundred miles, so the debris burned up or fell to Earth soon after. The Chinese strike left debris scattered at a five-hundred-mile altitude, where some sixty nations or government consortia today operate satellites.[53]

While nonstate terrorist groups are unlikely to acquire rockets capable of reaching the altitudes where GPS and GNSS satellites orbit, competition among nations that have heavy launchers and field GNSS satellites provides ample opportunities for miscalculations or accidents. Direct missile strikes pose less of a threat than other methods. China tested a microsatellite in late December 2008 that some analysts called a prototype for an antisatellite weapon. During a manned mission that included a spacewalk, the crew released a sixteen-inch cube weighing about ninety pounds that was capable of maneuvering around the spaceship. About four hours later the microsatellite flew within fifteen and a half miles of the International Space Station.[54] The Chinese said its purpose was to photograph and inspect their manned spacecraft. Many analysts noted similarities to co-orbital antisatellite designs dating back to the Cold War era. After the United States stopped overflying the Soviet Union in 1960 and switched to spy satellites the Soviets began their IS antisatellite program ("Istrebitel Sputnikov," or fighter satellite).[55] They designed these "kamikaze" satellites to carry explosives, launch like regular satellites, and maneuver near their targets to destroy them with shrapnel.[56] Operational versions of the system flew through 1983. The United States countered with Project SAINT (short for satellite interceptor), designed to inspect enemy satellites with television cameras, but canceled the program in 1963.[57] The United States conducted one test in 2005 of a satellite called the Demonstration for Autonomous Rendezvous Technology (DART), which featured similar capabilities.[58] A navigation error caused it to bump its target, a defunct communications satellite, into a different orbit without destruction or creation of debris.[59]

The Defense Department's *Annual Report to Congress: Military and Security Developments Involving the People's Republic of China* in May 2012 expressed unease about its microsatellite program. The report listed direct-ascent missiles, jamming, lasers, microwaves, and cyber weapons among China's counterspace capabilities and added, "Over the past two years, China has also conducted increasingly complex close proximity operations between satellites while offering little in the way of transparency or explanation."[60]

U.S. Navy commander John J. Klein, in *Space Warfare: Strategy, Principles and Policy*, hypothesizes that lesser space powers might contest command of space by locating microsatellites near GPS or GNSS satellites or even by attaching parasitic microsatellites to them, making defensive targeting difficult or impossible without destroying the navigation satellite itself.[61] A nation pursuing such a strategy need not arm the microsatellites with explosives; it might simply move them close enough to jam or block the transmitted signals or attach them physically to the satellite's antennas.[62] Deploying such microsatellites might gain their sponsor political influence without their ever being used. The strategy of using mere presence to influence policies of a dominant power is not new, Klein points out. The European Union achieved greater say in the allocation of radio spectrum by proposing to overlay GPS signals with Galileo.[63] Since Klein published his book, China demonstrated the same strategy by appropriating a frequency Galileo planned to use.

Lesser space powers are growing in number and capability and staking out their own presence in space. India successfully tested its Agni-V missile in April 2012, sending it to an altitude of about 373 miles. The head of India's Defence Research and Development Organization said the missile "ushered in fantastic opportunities in, say, building ASAT weapons and launching mini/micro satellites on demand."[64] He added that India did not plan to put weapons in space but had to "re-think" ASAT capability after China's 2007 demonstration.[65] Iran announced in June 2012 that it was completing work on a new space center at an undisclosed location that would accelerate domestic satellite launches.[66] Iran contracted with a Russian firm to design and launch its first satellite, Sinah-1, aboard a Kosmos rocket from northern Russia in 2005.[67] It launched its first homegrown satellite in 2009 and sent up another in 2010, along with a capsule carrying turtles, a rat, and a worm.[68] A 2011 attempt to launch a monkey into space failed, but Tehran claimed to have successfully orbited and returned a monkey to Earth in January 2013. That was widely declared a hoax after photographs of the monkey before and after the flight did not match. A February 2012 satellite launch was successful, but Iran delayed its remote-sensing Fajr satellite several times in 2012 and again in 2013. Some Western intelligence sources believe the delay announcements have been attempts to cover up launch failures.[69] It would be Iran's first satellite with maneuvering thrusters if it ever reaches space.[70] After several highly publicized failures, North Korea launched what it called an Earth observation satellite into orbit in December 2012. South Korean analysts who recovered the rocket booster reported evidence that the launch masked a military ICBM test, and astronomers said the satellite appeared dead.[71]

All of these activities lead some to conclude that an arms race is underway in space. Brian Weeden, technical advisor for the Secure World Foundation and a former Air Force captain at the U.S. Strategic Command's Joint Space Operations Center, notes that controlling an arms race in space is difficult because so many basic space technologies can be used for good or ill. For example, the same technology that enables automatic rendezvous and docking with the International Space Station could direct a co-orbital ASAT.[72] Dr. James Clay Moltz, a professor at the Naval Postgraduate School in the Department of National Security Affairs, describes four competing schools of thought driving U.S. space policy. They are space nationalism, which seeks dominant military power to control space; technological determinism, which advocates the selective military restraint of the Cold War; social interactionism, which emphasizes commercial cooperation; and global institutionalism, which favors international management regimes.[73] Which approach or combination of approaches will ultimately prevail is an open question.

Meanwhile space-tracking technology is on the rise. To overcome the limitations of monitoring orbital objects with ground-based radars and telescopes, the United States launched its first Space Based Space Surveillance (SBSS) satellite in September 2010.[74] Nearly two years later, in August 2012, the Air Force declared initial operational capability for the largely classified program.[75] The satellite, orbiting at 390 miles above Earth, and the ground control system together cost about $500 million.[76] Based on 2013 budget requests, the Air Force may be moving toward acquiring a second SBSS satellite.[77]

Observers have asked whether the U.S. military is too dependent on space, but when it comes to both military and civilian use of GPS, a common refrain is, "There is no going back."[78] That shared dependency comes with an enormous and growing financial commitment. The United States will spend nearly $10 billion for military space programs in fiscal year 2013 alone, but cost overruns can quickly eat up an entire year's worth of funding. In March 2012 GAO reported that spending on major defense satellite acquisition programs, of which GPS is but one, increased by about $11.6 billion—a 321 percent rise—from original cost estimates for the five-year fiscal period 2011 through 2016.[79] The first two GPS III satellites, at $1.6 billion and $1.4 billion respectively, did not meet the original schedule and exceeded their estimated cost by 18 percent.[80] As the United States grapples with long-term structural budget deficits, military budgets will come under pressure, and that will almost certainly affect the future of GPS.

What's Next for GPS Users?

Current applications are the starting point in forecasting how we may use GPS in coming decades. Military dependence on GPS may decrease as the Pentagon explores new techniques to replace, augment, or back up the signals. Often mentioned is developing eLoran, a modernized version of the ground-based long-range navigation system Loran C that the United States decommissioned in 2010. The enhanced version requires fewer ground stations and broadcasts high-powered, low frequency, unjammable signals for use on the earth's surface and up to jet altitudes.[81] The Defense Advanced Research Projects Agency (DARPA), through its Micro-PNT Program, is working on self-calibrating inertial navigation and guidance systems, which combine clocks, accelerometers, gyroscopes, and calibration circuitry into an eight-millimeter cube.[82] Self-contained inertial navigation units that do not require GPS to correct drift would be impervious to jamming or spoofing. A radio-navigation system that mimics GPS but works indoors and in underground mines, where GPS signals do not reach, has won Air Force backing. The Air Force signed a contract in 2010 with Locata, a privately owned Australian company, to provide a non-GPS component for its next-generation Ultra High Accuracy Reference System, an amalgamation of GPS and other PNT techniques aimed at overcoming jamming.[83]

Locata functions like a localized ground-based GPS without atomic clocks. A Locata network does not maintain absolute time based on an external standard. Rather, individual base stations, known as LocataLites, synchronize time with each other to within one or two nanoseconds.[84] An analogy would be a cappella singers matching pitch to each other rather than, say, a piano. Tests in October 2011 of an enhanced version of Locata's commercial off-the-shelf system confirmed that it met Air Force accuracy requirements. A plane flying at about 225 mph at an altitude of twenty-five thousand feet over an area roughly thirty by forty-five miles dotted with ten LocataLite antennas demonstrated horizontal accuracy of six centimeters (just over two inches) and vertical accuracy of fifteen centimeters (about six inches).[85] Locata's radio signals are more powerful than GPS, and they broadcast at the same frequency as Wi-Fi, making it easy to adapt many current devices to use them.[86] The Air Force signed a multiyear contract with Locata in September 2012 to install the system across a 2,500-square-mile area of the White Sands Missile Range, and a veteran Air Force GPS manager has joined the company.[87]

For the commercial sector and for consumers there are significant opportunities and challenges ahead in the area where most people first discovered GPS—surface transportation. Toll roads for years have used electronic transponders or bar codes to charge vehicles without stopping at tollbooths. GPS facilitates charging drivers by the mile, replacing gas-tax revenues lost to rising fuel efficiency. Switzerland implemented the first nationwide GPS-based toll system for trucks in 2003.[88] Germany followed two years later with a system covering 7,500 miles of the autobahn, and today a patchwork of satellite, cellular, and transponder systems covers Europe, spurring the development of hybrid electronic toll devices for vehicles.[89] Proponents of vehicle-miles-traveled (VMT) taxes in the United States face strong headwinds. A Congressional Budget Office study found some potential benefits in reducing emissions and addressing traffic congestion, but VMT taxes would be costly to implement and raise privacy issues.[90] Add to that the normal resistance to new taxes. When a North Carolina legislative committee floated the idea in 2009, a poll by the conservative Civitas Institute found 70 percent of voters against it.[91] Around the same time Transportation Secretary Ray LaHood, an Illinois Republican, called VMT an option "we should look at" in an Associated Press interview. White House press secretary Robert Gibbs dismissed the idea that same day.[92] Three years later the GOP campaign platform explicitly opposed "any funding mechanism that would involve governmental monitoring of every car and truck in the nation."[93] While a federal VMT seems unlikely, cash-strapped states are moving in that direction. California, Florida, Nevada, Minnesota, Oregon, and Washington are studying the approach.[94] One or more states appear likely to pass a VMT tax within five to ten years.[95]

Other GPS legislation may be forthcoming. Sen. Charles Schumer, a New York Democrat, has called for the Department of Transportation to set standards for information that GPS navigation systems must include, such as the height of overpasses.[96] Trucks have struck New York overpasses more than two hundred times since 2005. A quarter of those accidents occurred in Long Island alone, where one bridge has been hit twenty-seven times.[97]

A key development within sight is the fusion of GPS and other technologies used in machine control and transportation to create smarter, connected cars and intelligent transportation networks. Driver assistance technologies, such as backup cameras and parallel parking assist, and collision avoidance sensors, such as blind spot detection, adaptive headlights, and autonomous braking, have been available for years in luxury models. The 2012 model year brought a sudden expansion to a wide range of vehicles, according to ABI Research.[98] A Highway Data Loss Institute study released in mid-2012 showed that forward collision avoidance braking and headlights that shift toward curves reduced insurance claims by 14 percent.[99] Soon cars will not only give drivers more information and autonomously avoid collisions, they will interact with each other and with sensors and beacons built into highways. U.S. Department of Transportation officials believe vehicle-to-vehicle (V2V) and vehicle-to-infrastructure (V2I, or V2X) communication could prevent four out of five unimpaired vehicle crashes.[100] Motor vehicle accidents and fatalities have decreased slightly over the past decade, but in 2009 (the most recent figures available) there were 10.8 million motor vehicle accidents and 34,485 people died from motor-vehicle-traffic-related injuries, about a fifth of all injury deaths.[101] The National Highway Traffic Safety Administration and the University of Michigan's Transportation Research Institute in August 2012 launched the first large road test of connected vehicles. Three thousand cars, trucks, and buses equipped with devices to send and receive electronic Wi-Fi data messages tested the concept on the streets of Ann Arbor during the yearlong study.[102] NHTSA administrator David Strickland called V2V potentially "the ultimate game-changer in roadway safety."[103] A Ford Motor executive likened the technology to cars sending tweets to one another—short messages about speed, position, and direction—so even though a driver may not see another vehicle accelerating to beat a red light at a cross street, her car will warn her.[104] The government could mandate v2v technology in new cars as soon as 2017, though it would be years before most vehicles on the road have it.[105]

The possible changes wrought by two-way communications between vehicles and infrastructure are profound. Many traffic signals are already equipped to turn green when prompted by emergency vehicle transponders. Suppose traffic signals sent vehicles messages about the time remaining for a green light, as many crosswalk signs already do for pedestrians. Taken a step further, they could automatically alert or slow down vehicles if drivers failed to notice a red light. Many police departments now use portable digital signs with radar guns that flash the speed of oncoming cars, and most drivers voluntarily slow down. Once cars are suitably equipped, police could potentially use v2i communication to slow a vehicle down to the posted limit—or to send a citation.

Such communications do not require GPS, but some transportation theorists envision large-scale vehicular traffic management akin to air traffic control. A team of transportation and computer science researchers from Rutgers University, the University of California at Los Angeles, and the University of California at Berkeley has proposed a concept called "active highways" to manage traffic congestion the way computer networks handle large volumes of data.[106] Drivers enter the system using digital tickets, inform the system of travel plans, reserve slots in high-priority intelligent lanes, and comply with system adjustments designed to route vehicles around problems. This is a long way from a Sunday drive in the country. With such infrastructure in place, it is not hard to imagine eventually requiring drivers to relinquish control entirely in high-traffic areas where the highway management system and cars collaborate—eliminating unexpected and unsafe movements, speeding, tailgating, and accidents.

If it sounds farfetched, consider that by summer 2012 Google's experimental driverless cars surpassed three hundred thousand miles without an accident, a better record than the average driver, based on Federal Highway Administration statistics.[107] Nevada, Florida, and California already approved autonomous cars, and Google is spending millions lobbying other state and federal officials.[108] Governor Jerry Brown rode in a self-driving Prius to Google's Mountain View headquarters, where he signed the California law in late September 2012. Brown called the technology "science fiction becoming tomorrow's reality," and Google cofounder Sergey Brin told reporters, "Self-driving cars do not run red lights."[109] Google plans to sell its Driverless Car System, a technology package combining cameras, sensors, GPS, and software, to auto-makers after it wins certification from the National Transportation Safety Board and its international counterparts.[110] The first driverless cars could hit show-rooms within five years, and members of the Institute of Electrical and Elec-tronics Engineers (IEEE) predict that autonomous cars will make up three-quarters of those on the road by 2040.[111]

Some wonder if the changes in how we navigate are ruining our natural sense of direction or may cause unintended physiological or psychological side effects. As car navigation units became popular, contrarians noted that users were pay-ing more attention to their gadgets than the scenery.[112] Researchers at Cornell University studied drivers using GPS and found evidence of disengagement with the environment. Drivers became "immersed more in the virtual-technological environment," and some treated their GPS units like another occupant in the car, even giving them names. However, the researchers noted new opportunities for engagement, as drivers discovered landmarks otherwise invisible from the road but displayed in digital searches for particular points of interest, such as a type of restaurant.[113]

Brain research has established a connection between using spatial naviga-tional skills and gray-matter density of the hippocampus, a seahorse-shaped region of the brain associated with memory and spatial orientation.[114] Eleanor Maguire, a neuroscientist at University College London, found in a 2000 study that part of the hippocampus of London taxi drivers was larger than that of the general population, even bus drivers.[115] She linked the physical change to the cabbies' intimate knowledge of London's twenty-five thousand streets built up over years, as a muscle grows with use. Follow-up studies with brain scan-ners showed increased hippocampus activity as taxi drivers neared their des-tinations (driving cars in video games) but also revealed that a driver who suffered brain damage from a viral infection lost his superior ability to navigate the city's winding secondary streets.[116]

Muscles—and gray matter—atrophy with disuse, and the role of the hip-pocampus in memory and Alzheimer's disease has researchers concerned about GPS. Véronique Bohbot, a professor of psychiatry at McGill University and Douglas Institute in Montreal, focuses her brain research on human spa-tial memory, navigation, and plasticity. She has found that different naviga-tional strategies use different parts of the brain.[117] People who use spatial memory to form mental maps—learning the relationships between environ-mental landmarks—have more gray matter in the hippocampus. Following turn-by-turn directions—like those used in GPS—uses a different part of the brain. This means everyone is subject to the "use it or lose it" principle, not just taxi drivers.

Bohbot speculates that overreliance on GPS navigation could hasten normal age-related degeneration of the hippocampus and thereby increase the risk of dementia, previously associated with a reduction in volume in that part of the brain.[118] She recommends that people use GPS to learn about the environment but not become dependent on it. "Use it on your way to a new place, and use your memory on the way back," she counsels. "If you know that you will be turning off your GPS, you will pay attention to the environment, giving yourself a healthy cognitive workout."[119]

Technology, however, will not wait on the answers to such questions, and the trends for mobile geolocation applications and location-based services discussed in the preceding chapter suggest scenarios as futuristic as cars with autopilot. Google plans to solve the problem of people stumbling as they walk around staring down at smartphones with an interactive heads-up display built into a device worn like eyeglasses.[120] The company announced Project Glass in April 2012 and posted a concept video on YouTube demonstrating hands-free, voice-activated text messaging, video chat, pop-up maps, turn-by-turn directions, calendar reminders, music, check-in at locations via social networks, and photo taking—the majority of smartphone features.[121] The announcement sparked inevitable comparisons to science fiction cyborg characters such as the Terminator, RoboCop, and *Star Trek*'s Geordi Laforge. While Google refines the concept's implementation, it is working to enhance everyday appeal by teaming with fashion designer Diane von Furstenberg.[122]

Voice control has limitations, so some method of touch interaction seems probable. That could be a trackball on the device, a wirelessly connected smartphone, or a "smart glove" that allows users to "touch" the virtual screen projected in front of them to swipe an image or tap a keyboard.[123] The company is targeting the rechristened Google Glass for consumer release in 2014 at a price around $1,500, but details are subject to change—competition in the augmented-reality eyewear sector is likely to be fierce.[124] Apple has filed patents, and other companies pursing similar technology include video eyewear specialist Vuzix of Rochester, New York, multinational optical and imaging giant Olympus, the gaming software company Valve, and British-based TTP (The Technology Partnership).[125] Bellevue, Washington-based Innovega has created augmented reality contact lenses.[126] The company signed a contract in 2012 with DARPA to develop a system for warfighters with a heads-up display that superimposes battlefield data on the user's normal field of vision.[127]

How these technologies will evolve over the long term is anyone's guess. More than two hundred thousand people already have cochlear implants that transmit sounds from external microphones to their auditory nerves, and Boston neurosurgeons have wired an external lens directly to a blind man's optic nerve, allowing him to see colors and read large-print text.[128] Futuristic entertainment often presages real life, even when exaggerating dangers for dramatic effect. Warner Brothers in August 2012 released *H+: The Digital Series*, a collection of short-format dystopian science fiction episodes on YouTube. Humans in this near-future world have augmented reality neural implants, connecting them to the Internet 24/7 but also exposing them personally to computer viruses.[129] It is impossible to predict whether pop-up geolocation data—whether displayed in glasses, superimposed on our retinas, or mainlined into our cerebral cortex—may eventually do to street signs what Internet search has done to the Yellow Pages, but it seems likely that GPS will remain integral to communication, navigation, and augmented reality devices, whatever form they assume.

Looking back over the past half century of technological change it is clear that some things people assumed were inevitable have not happened and others emerged unexpectedly. In the early 1960s *The Jetsons* made youngsters believe flying cars would be commonplace by now.[130] In the late 1960s books like *The Population Bomb*, by Paul Ehrlich, made a case for imminent worldwide famine, but advances in food production derailed that prediction. Today GPS-enabled precision agriculture is leading another wave of productivity as its adoption spreads worldwide. During the early decades of development that led to global navigation satellite systems, many foresaw a new era of precision warfare. Some anticipated a spillover to civil aviation and ground transportation. A few envisioned worldwide-synchronized time and handheld devices for communication and navigation. Nobody predicted with clarity how putting GPS capability into the hands of individuals would spawn location-based services, enhanced social networks, or many other imaginative uses entrepreneurs have conceived.

As we approach the twentieth anniversary of GPS becoming operational, the pace of new application development, the diversity of uses, and the extent to which GPS-based technologies continue to revolutionize everyday activities suggest that human ingenuity is far from exhausting its potential benefits. However, we have no guidance system to chart where we will take the technology—or where GPS technology will take us.

Notes

Introduction

1. Except where noted, the entire account of this vehicle chase is drawn from Tom Krisher, "OnStar Shuts Down Visalia Carjacking in A Hurry," *San Francisco Chronicle*, October 20, 2009, http://articles.sfgate .com/2009-10-20/news/17186446_1_general-motors-co-s-onstar -onstar-service-high-speed-chase; Lewis Griswold, "OnStar Re-creates Visalia Rescue," *Fresno Bee*, November 19, 2009, http://www.fresnobee .com/columnists/griswold/story/1718168.html; General Motors, "OnStar Stolen Vehicle Slowdown Helps Recover Carjacked Vehicle, Prevents High-Speed Chase," news release, October 19, 2009, http:// publish.media.gm.com/content/media/us/en/news/news_detail .brand_gm.html/content/Pages/news/us/en/2009/Oct/1019_OnStar _Slowdown; Jocelyn Allen, OnStar spokesperson, telephone conversation with Eric Frazier, February 4, 2010.
2. General Motors, "OnStar Stolen Vehicle Slowdown Helps Recover Carjacked Vehicle."
3. *The Associated Press Stylebook and Briefing on Media Law, 2009* (Philadelphia PA: Basic Books, 2009), 122.
4. There is no consensus number for the total worldwide market due to differing methodologies used by market research companies. For

example, one 2009 report forecast world sales crossing $75 billion (in U.S. dollars) by 2013, while a 2011 report forecast a world market worth $26.67 billion in 2016. Reuters, "High Growth Reported for the World GPS Market Forecast to 2013," news release, April 23, 2009, http://www.reuters.com/article/2009/04/23/idus110111+23-Apr-2009+bw20090423.

5. PR Newswire, "MarketsandMarkets: Global GPS Market Worth $26.67 Billion by 2016," news release, November 24, 2011, http://media.prnewswire.com/en/jsp/latest.jsp?resourceid=4850435&access=EH; Amy Gilroy, "OnStar Generates More than $1B: iSuppli," *TWICE*, April 29, 2009, http://www.twice.com/article/235776-OnStar_Generates_More_Than_1B_iSuppli.php?rssid=20315&q=GPS.

6. GM News, "Onstar Empowers Everyday Citizens to Help Others," news release, March 12, 2013, http://media.gm.com/media/us/en/onstar/news.detail.html/content/Pages/news/us/en/2013/Mar/0312-good-samaritan.html; GM News, "OnStar RemoteLink Passes 14 Million Interactions," news release, May 23, 2012, http://media.gm.com/media/us/en/gm/news.detail.print.html/content/Pages/news/us/en/2012/May/0523_onstar.html.

7. Larry Copeland, "Technology May Halt Hot Pursuit; Many Hurt, Killed in Police Chases," *USA Today*, December 11, 2008, LexisNexis Academic.

8. General Motors, "OnStar Stolen Vehicle Slowdown Helps Recover Carjacked Vehicle."

9. Lou Prato, "Choppers Soar at Local News Operations," *American Journalism Review*, April 1997, http://www.ajr.org/Article.asp?id=347.

10. Copeland, "Technology May Halt Hot Pursuit."

11. Amy Gilroy, "Peering into the Future of the PND," *TWICE*, October 12, 2009, http://www.twice.com/article/print/357820-Peering_Into_The_Future_Of_The_PND.php.

12. CNN.com, "Transcript of *Fareed Zakaria GPS*, aired June 1, 2008, 13:00 ET," http://transcripts.cnn.com/TRANSCRIPTS/0806/01/fzgps.01.html (accessed February 16, 2010).

13. YouTube.com, "Jared Jewelry," http://www.youtube.com/watch?v=a69eC7ldAcI (accessed February 16, 2010).

14. YouTube.com, "TurboTax," http://www.youtube.com/watch?v=U7atn5Zj5hY (accessed February 16, 2010).

15. YouTube.com, "Fidelity—Turn Here," http://www.youtube.com/watch?v=U7atn5Zj5hY (accessed February 16, 2010).

16. Ki Mae Heussner, "The Top 10 Innovations of the Decade," ABCNews.com, http://abcnews.go.com/Technology/AheadoftheCurve/top-10-innovations-decade/story?id=9204931 (accessed December 15, 2009).

17. Rob Pegoraro, "A Tech Top 10 for the 2000s," *Faster Forward* (blog), WashingtonPost.com, December 14, 2009, http://voices.washingtonpost.com/fasterforward/2009/12/a_tech_top_ten_for_the_2000s.html (accessed December 22, 2009).

18. Nam D. Pham, *The Economic Benefits of Commercial GPS Use in the U.S. and the Costs of Potential Disruption* (Washington DC: NDP Consulting Group, June 2011), 3, http://www.saveourgps.org/pdf/GPS-Report-June-22-2011.pdf.

19. comScore, "comScore Reports January 2013 U.S. Smartphone Subscriber Market Share," news release, March 6, 2013, http://www.comscore.com/Insights/Press_Releases/2013/3/comScore_Reports_January_2013_U.S._Smartphone_Subscriber_Market_Share.

20. Brian X. Chen, "Get Ready for 1 Billion Smartphones by 2016, Forrester Says," *Bits* (blog), *New York Times*, February 14, 2012, http://bits.blogs.nytimes.com/2012/02/13/get-ready-for-1-billion-smartphones-by-2016-forrester-says; Berg Insight, "Global PND Shipments Declined to 33 Million Units in 2012," news release, totaltele.com, January 18, 2012, LexisNexis Academic.

21. Berg Insight, "Global PND Shipments Declined to 33 Million Units"; Business Wire, "PND Vendors Shifting Strategies to Deal with Competitive Pressures from Mobile Apps, Says ABI Research," news release, March 8, 2012, LexisNexis Academic.

22. GPS.gov, "Applications: Timing," http://www.gps.gov/applications/timing (accessed August 5, 2012).

23. Pham, *Economic Benefits of Commercial GPS Use*, 1.

24. National Executive Committee for Space-Based PNT, "Charter," http://www.pnt.gov/charter (accessed August 5, 2012).

25. GPS.gov, "National Space Policy, June 28, 2010, Excerpt," http://www.gps.gov/policy/docs/2010 (accessed August 10, 2012).

26. Scott Pace, "PNT Evolution: Future Benefits and Policy Issues," presentation to National Space-Based PNT Advisory Board, November 5, 2009, Alexandria VA, http://pnt.gov/advisory/2009/11/pace.pdf.

1. New Moons Rising

1. YouTube.com, "Vanguard TV3 Failed Rocket Launch," March 10, 2010, http://www.youtube.com/watch?v=zVeFkakURXM.

2. Associated Press, "Launching Hour Nears for Satellite," *Ocala (FL) Star-Banner*, December 4, 1957, http://news.google.com/newspapers?id=P_0jAAAAIBAJ&sjid=2QQEAAAAIBAJ&pg=4970,3889759&dq=december+6+1957+vanguard&hl=en.

3. Matt Bille and Erika Lishock, *The First Space Race: Launching the World's First Satellites* (College Station TX: Texas A&M University Press, 2004), ix.

4. Bille and Lishock, *First Space Race*, 123.

5. Martin Votaw, conversation with Eric Frazier, Hilton Alexandria Old Town, Alexandria VA, April 1, 2010.

6. Votaw, conversation with Eric Frazier.

7. Roger Easton, telephone conversation with Eric Frazier, March 14, 2010.

8. Roger Easton and Ruth Easton, conversation with Eric Frazier, Hilton Alexandria Old Town, Alexandria VA, April 1, 2010.

9. National Air and Space Museum, "50 Years of the Space Age: 1957: Vanguard TV3 Satellite," November 6, 2011, http://www.nasm.si.edu/collections/artifact.cfm?id=a19761857000.

10. Milton Bracker, "Vanguard Rocket Burns on Beach," *New York Times*, December 7, 1957, http://select.nytimes.com/gst/abstract.html?res=F50A13FC395E147B93C5A91789D95F438585F9.

11. "Vanguard's Aftermath: Jeers and Tears," *Time*, December 16, 1957, http://timedemo.newscred.com/article/8283732ce6e174a01ff40fb23be484e6.html/edit.

12. "Vanguard's Aftermath: Jeers and Tears."

13. Bille and Lishock, *First Space Race*, 123.

14. Constance McLaughlin Green and Milton Lomask, "From Sputnik I to TV-3," in *Vanguard: A History* (Washington DC: NASA, 1970), http://www.spacearium.com/filemgmt_data/files/Vanguard_A_History.pdf.

15. Drew Pearson, "Ike Said to Prefer Dying with His Boots on Rather Than Resign," *St. Petersburg (FL) Times*, December 14, 1957.

16. Robert Dallek, *Lyndon B. Johnson: Portrait of a President* (New York: Oxford University Press, 2004), 110.

17. Pew Research Center for the People and the Press, *Majority Says the Federal Government Threatens Their Personal Rights* (Washington DC, January 31, 2013), 6, http://www.people-press.org/files/legacy-pdf/01-31-13%20Views%20of%20Government.pdf; Pew Research Center for the People and the Press, *Distrust, Discontent, Anger and Partisan Rancor: The People and Their Government* (Washington DC, April 18, 2010), 13, http://people-press.org/reports/pdf/606.pdf.

18. Nancy E. Bernhard, *U.S. Television News and Cold War Propaganda, 1947–1960* (New York: Cambridge University Press, 1999), 47.

19. Bernhard, *U.S. Television News and Cold War Propaganda*, 47; U.S. Department of Commerce, Bureau of the Census, "Table No. 46—Households, by Race of Head, with Population per Household," in *Statistical Abstract of the United States, 1958* (Washington DC: U.S. Government Printing Office, 1958), 45.

20. Andrew J. Dunar, *America in the Fifties* (Syracuse NY: Syracuse University Press, 2006), 237.

21. Dunar, *America in the Fifties*, 248–49.

22. Dunar, *America in the Fifties*, 249.

23. "Satellite a Bust. Rocket Blows Up in First U.S. Try," December 9, 1957, in Universal Newsreels collection, Internet Archive, http://www.archive.org/details/1957-12-09_Satellite_A_Bust.

24. Sandy Stiles, "As Impatience Mounts, Fidgety Scientists Fuss with Bride-Like Missile," *St. Petersburg (FL) Times*, December 5, 1957.

25. Sandy Stiles, "Press Corps Stands, Waits for Launching," *St. Petersburg (FL) Times*, December 5, 1957.

26. Stiles, "Press Corps Stands."

27. "Vanguard's Aftermath: Jeers and Tears."

28. "The Press: Monday-Morning Missilemen," *Time*, December 23, 1957, www.time.com/time/magazine/article/0,9171,936782,00.html.

29. Associated Press, "Launching Hour Nears For Satellite."

30. Associated Press, "Launching Hour Nears For Satellite."

31. Project Vanguard Staff, *Project Vanguard Report No. 1: Plans, Procedures, and Progress*, NRL Report 4700 (Washington DC: Naval Research Laboratory, January 13, 1956), 28.

32. Project Vanguard Staff, *Project Vanguard Report No. 1*, 28.

33. Green and Lomask, "Early Test Firings," in *Vanguard: A History*.

34. Mike Gruntman, *Blazing the Trail: The Early History of Spacecraft and*

Rocketry (Reston VA: American Institute of Aeronautics and Astronautics, 2004), 373.

35. Paul Dickson, *Sputnik: The Shock of the Century* (New York: Walker Publishing, 2001), 76–77.

36. Dickson, *Sputnik*, 76–77.

37. National Academies, "The International Geophysical Year," March 16, 2010, http://www.nationalacademies.org/history/igy/.

38. United Press, "First Man-Made Moon May Be Visible in 1957," *Sarasota (FL) Herald-Tribune*, February 15, 1957, http://news.google.com/newspapers?nid=1755&dat=19560214&id=miEhAAAAIBAJ&sjid=7mQEAAAAIBAJ&pg=2518,3926077.

39. United Press, "First Man-Made Moon May Be Visible."

40. Dickson, *Sputnik*, 83.

41. Naval Research Laboratory, Rocket Development Branch, *A Scientific Satellite Program*, NRL Memorandum Report 466 (Washington DC: Naval Research Laboratory, April 13, 1955), 2.

42. Naval Research Laboratory, Rocket Development Branch, *Scientific Satellite Program*, 2.

43. Milton W. Rosen, *The Viking Rocket Story* (New York: Harper & Brothers, 1955), 232.

44. Rosen, *Viking Rocket Story*, 17.

45. Naval Research Laboratory, "Viking Program," http://www.nrl.navy.mil/accomplishments/rockets/viking-program (accessed July 31, 2012).

46. Roger Easton, interview by Dr. David van Keuren and James Tugman, Naval Research Laboratory History Office, Washington DC, March 8, 1996.

47. Green and Lomask, "Seeking Government Support for a Satellite Program," in *Vanguard: A History*.

48. John T. Mengel and Paul Herget, "Tracking Satellites by Radio," *Scientific American*, January 1958, 23–29.

49. C. A. Schroeder, C. H. Looney Jr., and H. E. Carpenter Jr., *Project Vanguard Report No. 18: Minitrack Report No. 1—Phase Measurement* (Washington DC: Naval Research Laboratory, July 26, 1957), 15.

50. Naval Research Laboratory, Rocket Development Branch, *Scientific Satellite Program*, 2–3.

51. Naval Research Laboratory, Rocket Development Branch, *Scientific Satellite Program*, 5–6.

52. Naval Research Laboratory, Rocket Development Branch, *Scientific Satellite Program*, 1.

53. Naval Research Laboratory, Rocket Development Branch, *Scientific Satellite Program*, 1.

54. National Security Council, "NSC 5520 Draft Statement of Policy on U.S. Scientific Satellite Program," May 20, 1955, http://www.thespacereview .com/archive/995b.pdf.

55. Homer J. Stewart, interview by John L. Greenberg, Pasadena CA, October–November 1982, and interview by Shirley K. Cohen, Altadena CA, November 3, 1993, Oral History Project, California Institute of Technology Archives, http://resolver.caltech.edu/CaltechOH:OH _Stewart_H (accessed March 12, 2010).

56. Green and Lomask, "Seeking Government Support."

57. White House, "Statement by James C. Hagerty," news release, July 29, 1955, Dwight D. Eisenhower Presidential Library, Abilene KS, http:// www.eisenhower.archives.gov/research/online_documents/igy/1955_7 _29_Press_Release.pdf (accessed April 26, 2010).

58. Green and Lomask, "Selecting a Satellite Plan," in *Vanguard: A History*.

59. Clifford C. Furnas, "Why Did U.S. Lose the Race? Critics Speak Up," *Life*, October 21, 1957, 22.

60. Green and Lomask, "Selecting a Satellite Plan."

61. Dickson, *Sputnik*, 261; Green and Lomask, "Selecting a Satellite Plan."

62. Furnas, "Why Did U.S. Lose the Race?"

63. Dickson, *Sputnik*, 85.

64. "Memorandum of Conference with the President," October 9, 1957, Dwight D. Eisenhower Presidential Library, Abilene KS, http://www .eisenhower.archives.gov/research/online_documents/sputnik/10_9 _57_Early_Memo.pdf (accessed April 26, 2010).

65. National Museum of the Air Force, "Open Skies Proposal," December 6, 2006, http://www.nationalmuseum.af.mil/factsheets/factsheet.asp ?id=1878.

66. National Museum of the Air Force, "Open Skies Proposal."

67. National Security Council, "NSC 5520 Draft Statement of Policy."

68. Stephen E. Ambrose, *Eisenhower: Soldier and President* (New York: Simon & Schuster, 1990), 450.

69. Dwight D. Eisenhower, "Press Conference #123," October 9, 1957, transcript, Presidential Papers, Press Conference Series, Box 6,

Dwight D. Eisenhower Presidential Library, Abilene KS, http://www
.eisenhower.archives.gov/research/online_documents/sputnik/10_9
_57.pdf (accessed April 26, 2010).

70. Eisenhower, "Press Conference #123."

71. Dickson, *Sputnik*, 148.

72. Gruntman, *Blazing the Trail*, 359–60.

73. Dickson, *Sputnik*, 87–89.

74. Dickson, *Sputnik*, 89.

75. Roger Easton, e-mail to Al Nagy et al., March 29, 2008.

76. Dickson, *Sputnik*, 91–92.

77. NASA.gov, "Explorer-I and Jupiter-C: The First United States Satellite
and Launch Vehicle," http://history.nasa.gov/sputnik/expinfo.html
(accessed August 1, 2012).

78. Martin Votaw, "50th Anniversary Celebration of Vanguard One,"
unpublished transcript of panel discussion, Naval Research Laboratory, Washington DC, March 17, 2008; Gruntman, *Blazing the Trail*,
365.

79. Army.mil, "Remarks of the Rev. J. Bruce Medaris (MG, U.S. Army
Retired), 20th Anniversary of Explorer 1, Huntsville–Madison County
Chamber of Commerce, January 31, 1978," http://www.redstone.army
.mil/history/medaris/remarks.html (accessed April 26, 2010).

80. Matthew Brzezinski, *Red Moon Rising: Sputnik and the Hidden Rivalries
that Ignited the Space Age* (New York: Times Books, 2007), 196.

81. William J. Jorden, "'Mighty' Hydrogen Device Is Tested by Soviet
Union; Presumably in Central Asia," *New York Times*, October 8, 1957,
http://select.nytimes.com/gst/abstract.html?res=FB0812F63B5C127A
93CAA9178BD95F438585F9.

82. Sally Rosen, conversation with Richard Easton, June 16, 2009,
Bethesda MD.

83. Harold Callender, "Europeans Weigh Threats to World Peace,"
December 23, 1956, http://select.nytimes.com/gst/abstract.html?res
=F4071EFD3B5A137A93C1AB1789D95F428585F9.

84. "Man of the Year: Up from the Plenum," *Time*, January 6, 1958, http://
www.time.com/time/magazine/article/0,9171,868089,00.html.

85. C. D. Jackson memorandum to Henry R. Luce, October 8, 1957, in C. D.
Jackson Papers, Box 69, Log-1957 (4), Dwight D. Eisenhower Presidential Library, Abilene KS, http://www.eisenhower.archives.gov/

Research/Digital_Documents/Sputnik/10-8-57_Memo.pdf (accessed April 26, 2010).

86. White House Staff 1953–1961, "C.D. Jackson," Dwight D. Eisenhower Presidential Library, Abilene KS, http://www.eisenhower.archives.gov/All_About_Ike/Presidential/White_House_Staff_1953_1961/pages/WH%20staff%20book_031_0001_jpg.htm (accessed April 26, 2010).

87. Jackson memorandum to Luce.

88. Eisenhower, "Press Conference #123."

89. Stiles, "Press Corps Stands."

90. Dickson, *Sputnik*, 141–45.

91. Bille and Lishock, *First Space Race*, 122.

92. Project Vanguard Staff, *Project Vanguard Report No.1*, 28, 60.

93. Project Vanguard Staff, *Project Vanguard Report No. 9: Progress through September 15, 1956*, NRL Report 4850 (Washington DC: Naval Research Laboratory, October 4, 1956), 2.

94. Project Vanguard Staff, *Project Vanguard Report No. 20: Progress through July 31, 1957*, NRL Report 5020 (Washington DC: Naval Research Laboratory, September 16, 1957), 1.

95. Project Vanguard Staff, *Project Vanguard Report No. 20*.

96. Gruntman, *Blazing the Trail*, 358.

97. Roger Easton, "50th Anniversary Celebration of Vanguard One," unpublished transcript, Naval Research Laboratory, Washington DC, March 17, 2008.

98. Project Vanguard Staff, *Project Vanguard Report No. 20*, 1.

99. Votaw, "50th Anniversary Celebration of Vanguard One."

100. Votaw, "50th Anniversary Celebration of Vanguard One."

101. This entire summary of the fabrication of the satellite is based on Votaw's recollections in "50th Anniversary Celebration of Vanguard One."

102. "Vanguard's Aftermath: Jeers and Tears."

103. Green and Lomask, "From Sputnik I to TV-3."

104. NASA.gov, "Sputnik 1," National Space Science Data Center ID: 1957-001B, http://nssdc.gsfc.nasa.gov/nmc/spacecraftDisplay.do?id=1957-001B (accessed May 12, 2010).

105. NASA.gov, "Sputnik 2," National Space Science Data Center ID: 1957-002A, http://nssdc.gsfc.nasa.gov/nmc/masterCatalog.do?sc=1957-002A (accessed May 12, 2010).

106. NASA.gov, "Explorer: America's First Spacecraft: Explorer 1 Overview," http://www.nasa.gov/mission_pages/explorer/explorer-overview.html (accessed May 12, 2010).

107. Naval Research Laboratory, "Vanguard Project," http://www.nrl.navy.mil/accomplishments/rockets/vanguard-project/ (accessed July 31, 2012).

108. Green and Lomask, "The National Academy of Sciences and the Scientific Harvest, 1957–1959," in *Vanguard: A History*.

109. Project Vanguard Staff, *Project Vanguard Report No. 30: Progress through December 31, 1957*, NRL Report 5113 (Washington DC: Naval Research Laboratory, March 27, 1958), 15.

110. Project Vanguard Staff, *Project Vanguard Report No. 30*, 16.

111. William H. Guier and George C. Weiffenbach, "Genesis of Satellite Navigation," *Johns Hopkins APL Technical Digest* 19, no. 1 (1998): 14–17, www.jhuapl.edu/techdigest/td/td1901/guier.pdf.

112. Guier and Weiffenbach, "Genesis of Satellite Navigation."

113. Johns Hopkins University Applied Physics Laboratory, "An Overview of the Navy Navigation Satellite System," http://sd-www.jhuapl.edu/Transit/ (accessed May 10, 2010).

114. Johns Hopkins University Applied Physics Laboratory, "Overview of the Navy Navigation Satellite System."

115. Green and Lomask, "Success—and After," in *Vanguard: A History*.

116. NASA.gov, "History of NASA GSFC Tracking Services," http://esc.gsfc.nasa.gov/157.html (accessed July 31, 2012).

117. NASA.gov, "History of NASA GSFC Tracking Services."

118. Easton, interview by van Keuren and Tugman.

119. Easton, interview by van Keuren and Tugman.

120. U.S. Space Command, "Space Surveillance Network," http://www.au.af.mil/au/awc/awcgate/usspc-fs/space.htm (accessed July 31. 2012).

121. Easton, interview by van Keuren and Tugman.

122. Easton, interview by van Keuren and Tugman.

123. Navy.mil, "Navy Transfers Space Surveillance Mission to Air Force," news release, October 20, 2004, http://www.navy.mil/submit/display.asp?story_id=15597.

124. Roger Easton, "In the Beginning of GPS," keynote address, Thirty-Second Annual Precise Time and Time Interval (PTTI) Meeting, Hyatt Regency Hotel, Reston VA, November 28, 2000.

125. Easton, "In the Beginning of GPS."

2. Weather Permitting

1. Sarepta.org, "Navigation Satellites," http://www.sarepta.org/en/objekt .php?aid=209&bid=246&oid=1548 (accessed September 9, 2012).

2. Caroline Alexander, *The Endurance: Shackleton's Legendary Antarctic Expedition* (New York: Alfred A. Knopf, 1999), 150.

3. Mark Denny, *The Science of Navigation: From Dead Reckoning to GPS* (Baltimore MD: Johns Hopkins University Press, 2012), 45–47.

4. Torben B. Larsen, "Siwa: Oasis Extraordinary," *Saudi Aramco World*, September–October 1988, 2–7, http://www.saudiaramcoworld.com/ issue/198805/siwa-oasis.extraordinary.htm.

5. Pat Norris, "The Longitude Challenge—Apollo 8," *Navigation News: The Magazine of the Royal Institute of Navigation*, March–April 2012, 16–18.

6. Heloise Finch-Boyer, "Chart Wars," presentation at After Longitude—Modern Navigation in Context conference, National Maritime Museum, Greenwich, England, March 22, 2012, http://www.rin.org.uk/ Uploadedpdfs/ItemAttachments/Heloise%20Finch-Boyer%20-%20 presentation-web.pdf.

7. Daniel Gallery, *Cap'n Fatso* (New York: W. W. Norton, 1969), 119.

8. Amir Aczel, *The Riddle of the Compass* (New York: Harcourt, 2001), 11.

9. Aczel, *Riddle of the Compass*, 12–16.

10. Astrolabes.org, "The Astrolabe: An Instrument with a Past and a Future," http://www.astrolabes.org (accessed June 20, 2010).

11. Jeanne Willoz-Egnor, "Celestial Navigation Instruments: Cross-Staff," Institute of Navigation, Navigation Museum, http://www.ion.org/ museum/item_view.cfm?cid=6&scid=13&iid=26 (accessed August 8, 2012); Jeanne Willoz-Egnor, "Celestial Navigation Instruments: Octant," Institute of Navigation, Navigation Museum, http://www.ion .org/museum/item_view.cfm?cid=6&scid=13&iid=27 (accessed August 8, 2012).

12. Mariners' Museum, "Beveled Scale Sextant," http://www.mariner.org/ collections/beveled-scale-sextant (accessed June 20, 2010).

13. Trond Austheim and Børge Solem, "The S/S Atlantic of the White Star Line, Disaster in 1873," NorwayHeritage.com, http://www.norway heritage.com/articles/templates/great-disasters.asp?articleid=1 &zoneid=1 (updated May 2006); Nova Scotia's Electric Scrapbook, "S.S. Atlantic," http://ns1763.ca/hfxrm/ssatlansos.html (accessed June

14, 2011); SSAtlantic.com, "Rev. Ancient's Account," http://www
.ssatlantic.com/ancients.shtml (accessed October 27, 2004).

14. Joseph Patrick Bulko, "The Story of Weems & Plath," *Inside Annapolis*,
June–July 2007, http://www.insideannapolis.com/archive/2007/
issue4/weems.html.

15. N. W. Emmot, "The Grand Old Man of Navigation," *Quarterly Newsletter of the Institute of Navigation* 16, no. 3 (Fall 2006): 4–7, www.ion.org/
newsletter/v16n3.pdf. See also Roger Connor, "Even Lindbergh Got
Lost," *Air and Space Magazine*, February/March 2013, 28–33.

16. Alexander H. Flax, e-mail to Richard Easton, September 2, 2012.

17. Michael Russell Rip and James M. Hasik, *The Precision Revolution:* GPS
and the Future of Aerial Warfare (Annapolis MD: Naval Institute Press,
2002), 19.

18. Denny, *Science of Navigation*, 213.

3. Success Has Many Fathers

1. Arthur C. Clarke, letter to Andrew G. Haley, August 5, 1956, reproduced at TechJournal.com, "Sci-Fi Author Arthur C. Clarke Predicted
GPS, Satellite TV and Cellphones," April 28, 2012, http://thetechjournal
.com/off-topic/sci-fi-author-arthur-c-clarke-predicted-gps-satellite-tv
-and-cellphones.xhtml.

2. Roy E. Anderson, "General Electric: Early Space Age Adventures,"
Quest: The History of Spaceflight Quarterly 15, no. 3 (2008): 53.

3. Anderson, "General Electric," 53.

4. Roy E. Anderson to Eric Frazier, July 1, 2011, copy in Frazier's files.

5. R. E. Anderson, "A Navigation System Using Range Measurements
with Cooperating Ground Stations," *Navigation: Journal of the Institute
of Navigation* 2, no. 2 (Autumn 1964): 319.

6. Philip Klass (*Aviation Week & Space Technology* senior editor) to Howard
Marx, December 26, 1973, copy in Roger Easton's files: "For example,
the idea of one-way (passive) ranging was proposed by The National
Corp., then a producer of atomic clocks, for possible use in mid-air collision avoidance systems back in the mid-fifties (as I recall) and was
tested, under FAA sponsorship, in the late 1950s. The idea even then
was under investigation by Sierra Research for military uses."

7. Harold Rosen, conversation with Richard Easton, May 29, 2012. See
also Harold A. Rosen, "Syncom: World's First Geostationary Satellite,"

in *Success Stories in Satellite Systems*, ed. D. K. Sachdev (Reston VA: American Institute of Aeronautics and Astronautics, 2009), 19–20.

8. NASA.gov, "What is a Sounding Rocket," April 12, 2004, http://www .nasa.gov/missions/research/f_sounding.html.

9. Naval Research Laboratory, Rocket Development Branch, *Scientific Satellite Program*, 10.

10. FAS.org, "U.S. Naval Space Command Space Surveillance System," http://www.fas.org/spp/military/program/track/spasur_at.htm (accessed April 2, 2010).

11. Martin Votaw, conversation with Richard Easton, September 12, 2009.

12. Rick Sturdevant, "Tracing Connections—Vanguard to NAVSPASUR to GPS: An Interview with Roger Easton," *High Frontier: The Journal for Space and Missile Professionals* 4, no. 3 (May 2008): 53.

13. Easton, interview by van Keuren and Tugman.

14. Chester Kleczek, interview by Richard Easton, June 19, 2009. Kleczek stated that electronically steerable radars have limitations since they do not cover much area. The continuous wave that Space Surveillance used was superior.

15. Peter Wilhelm, conversation with Richard Easton, March 31, 2010.

16. R. L. Easton, *Space Applications Branch Technical Memorandum No. 1: An Exploratory Development Program in Passive Satellite Navigation* (Washington DC: Naval Research Laboratory, May 8, 1967), 1.

17. Naval Research Laboratory, "Historical Data," unattributed Timation chronology, July 22, 1971.

18. H. M. Smith, N. P. J. O'Hora, R. Easton, J. Buisson, and T. McCaskill, "International Time Transfer between USNO and RGO Via NTS-1 Satellite," paper presented at the Seventh Annual Precise Time and Time Interval (PTTI) Meeting, Greenbelt MD, December 1975, 341, http:// www.pttimeeting.org/archivemeetings/1975papers/Vol%2007_18.pdf.

19. Robert R. Whitlock and Thomas B. McCaskill, *NRL GPS Bibliography: An Annotated Bibliography of the Origin and Development of the Global Positioning System at the Naval Research Laboratory* (Washington DC: Naval Research Laboratory, June 3, 2009), 9.

20. Easton, interview by van Keuren and Tugman; Chester Kleczek, conversation with Richard Easton, March 31, 2010.

21. Easton, interview by van Keuren and Tugman.

22. Kleczek, interview by Easton.

23. Kleczek, interview by Easton.

24. The July 22, 1971, chronology (Naval Research Laboratory, "Historical Data") called this change in navigation technique the "celestial transformation."

25. Thomas McCaskill and Robert Whitlock, "40th Anniversary of Historic NRL Navigation Demonstration," *Labstracts*, October 15, 2007, 6.

26. James Buisson, e-mail to Richard Easton, August 3, 2009.

27. Naval Research Laboratory, *Timation Development Plan*, NRL Report 7227, rev. ed. (Washington DC, March 2, 1971), 1.

28. Don Jewell, "GPS Insights—January 2008," *GPS World*, January 2008, http://www.gpsworld.com/defense/gps-insights-january-2008-8438.

29. M. P. Gleason, "Galileo: Power, Pride, and Profit" (PhD diss., George Washington University, 2009), 5.

30. Anderson, "Navigation System Using Range Measurements," 317.

31. Easton, *Space Applications Branch Technical Memorandum No. 1*, 3.

32. Jonathan Eberhart, "The Time Web," *Science News*, July 13, 1974, 27.

33. Whitlock and McCaskill, NRL GPS *Bibliography*, 11.

34. J. B. Woodford and H. Nakamura, *Briefing-Navigation Satellite Study* (Los Angeles: Air Systems Command, August 24, 1966), 7; Ivan Getting "The Global Positioning System," IEEE *Spectrum*, December 1993, 44. The study chronology shows that Ivan Getting's comment in his IEEE *Spectrum* article that the Air Force's satellite work "stimulated additional satellite-based position-location and navigation by the Navy" is incorrect because the Navy's work began contemporaneously and neither service was initially aware of the other's projects.

35. Woodford and Nakamura, *Briefing-Navigation Satellite Study*; Getting, "Global Positioning System," 56.

36. J. B. Woodford, W. C. Melton, and R. L. Dutcher, "Satellite Systems for Navigation Using 24-Hour Orbits," paper presented at Electronics and Aerospace Systems Convention (EASCON), Washington DC, 1969, 184.

37. *Encyclopedia of Science*, s.v. "SECOR," http://www.daviddarling.info/encyclopedia/S/SECOR.html (accessed July 29, 2011).

4. One System, Two Narratives

1. Old Bailey Proceedings Online, "December 1692 Trial of John Glendon," June 12, 2011, http://www.oldbaileyonline.org/browse.jsp?foo =bar&path=sessionsPapers/16921207.xml&div=t16921207-7.

2. Bradford Parkinson, "GPS for Humanity," lecture, Stanford University, Stanford CA, April 30, 2012, http://www.youtube.com/watch?v =d6i6wFf-X_c (accessed July 8, 2012); Edward H. Martin, "The Early Days [of GPS]," lecture, Ohio State University, March 17, 2011, http://www.gps.gov/multimedia/presentations/2011/03/ohio/martin.pdf (accessed July 8, 2012). Martin did not mention Timation, and his only mention of the Naval Research Laboratory and Roger Easton referred to the NRL's role as a new customer for Magnavox Research Labs.

3. U.S. General Accounting Office, *Joint Major System Acquisition by the Military Services: An Elusive Strategy*, GAO/NSIAD-84-22 (Washington DC, December 23, 1983).

4. Harry Sonnemann, memorandum to Eric Frazier, April 30, 2010.

5. Ron Beard, letter to the editor, *GPS World*, December 2010.

6. Tom McCaskill, conversation with Richard Easton, April 28, 2007.

7. Harry Sonnemann, memorandum to Roger Easton, August 28, 1994.

8. Rick W. Sturdevant, "NAVSTAR, the Global Positioning System: A Sampling of its Military, Civil, and Commercial Impact," in *Societal Impact of Spaceflight*, ed. Stephen J. Dick and Roger D. Launius (Washington DC: NASA, 2007), 331.

9. Sonnemann, memorandum to Frazier.

10. Ron Beard, letter to the editor, *GPS World*, December 2010.

11. Keith D. McDonald, "Global Positioning System: Origins, Early Concepts, Development, and Design Success," in Sachdev, *Success Stories in Satellite Systems*, 250.

12. *Timation Development Plan*, 10.

13. Bradford W. Parkinson and Stephen W. Gilbert, "NAVSTAR: Global Positioning System—Ten Years Later," *Proceedings of the IEEE* 71, no. 10 (October 1983): 1177: "The Timation concept was essentially a two-dimensional system and lacked the ability to provide continuous position updates in a high-dynamic aircraft environment." Bradford W. Parkinson, letter to the editor, *Inside GNSS*, May 2010: "His method was obviously two-dimensional (called for 'Lines of Position'). Had he tried for more, his signal would have interfered with itself." Bradford W. Parkinson and Stephen T. Powers, "Fighting to Survive: Five Challenges, One Key Technology, the Political Battlefield—and a GPS Mafia," *GPS World*, June 2010, 14: "Both the description and the accompanying diagram in the patent clearly refer to two-dimensional

navigation, using lines of position." Art McCoubrey, who with Bob Kern designed the first cesium atomic clocks in orbit, launched in NTS-2 in 1977, in an August 26, 2011, e-mail to Richard Easton made the apt comment, "Dimensionality of Timation: I do not remember any discussions of dimensionality. I think that I may have regarded this issue as not very important because, given enough viewable satellites in the system, I thought it should be possible to get all of the position information as well as the system time." Speculation about side-tone signal interference has obscured the fact that Timation plans also specified using spread spectrum signals. See also Whitlock and McCaskill, *NRL GPS Bibliography*, 29: "NRL formulated and implemented a solution to the problem of instantaneous navigation using artificial satellites by passive ranging." The Navy already had a two-dimensional space-based navigation system in Transit. It made no sense to build another two-dimensional system.

14. Roger Easton, "Timation Navigation Satellite," in *Proceedings of the Second Precise Time and Time Interval (PTTI) Strategic Planning Meeting* (Washington DC: Naval Research Laboratory, December 10–11, 1970), 2:15.

15. Bob Kern, memorandum to Art McCoubrey, September 27, 1972. Kern mentioned that Easton was "ecstatic about the Efratom performance during a centrifuge test that went as far as 10 Gs. Roger is obviously abandoning his favoritism towards quartz and is talking about atomic resonances for use in satellites."

16. David C. Holmes, interview by John H. Bryant, July 13, 1994, copy of transcript of unpublished IEEE interview in the possession of the authors.

17. Brad Parkinson, "True Origins and Major Original Challenges for GPS Success, 1962–1978," PowerPoint presentation, October 2009, slide 25, http://scpnt.stanford.edu/pnt/PNT09/presentation_slides/13 _Parkinson_Creating_GPS.pdf (accessed June16, 2013) mentions that at Lonely Halls, only JPO and Aerospace personnel were in attendance, about twelve in total. Attendees included Parkinson, Maj. Gaylord Green, and Maj. Mel Birnbaum. In his 2012 presentation "GPS for Humanity," Parkinson added Frank Butterfield of Aerospace, Bill Huston of the Navy, and Steve Gilbert of the Air Force to the list of people at Lonely Halls. However, Bill Huston was appointed as Navy deputy program manager to the GPS JPO on January 15, 1974. Thus, he was not officially part of the JPO in September 1973.

18. Roger Easton, "Memoir," March 29, 2005, http://www.gpsinventor.com/?page=memoir.

19. Ron Beard, conversation with Richard Easton, September 2, 2011.

20. Ron Beard, e-mail to Richard Easton, August 4, 2011.

21. Sonnemann, memorandum to Frazier.

22. McDonald, "Global Positioning System," 241; Keith D. McDonald, telephone interview by Eric Frazier, July 23, 2010.

23. Parkinson and Gilbert, "NAVSTAR: Global Positioning System."

24. Brad Parkinson, oral history conducted November 2, 1999, by Michael Geselowitz, IEEE History Center, New Brunswick NJ, http://www.ieeeghn.org/wiki/index.php/OralHistory:Brad_Parkinson.

25. David C. Holmes, "NAVSTAR Global Positioning System: Navigation for the Future," *Proceedings* (U.S. Naval Institute), April 1977, 101.

26. Phillip J. Klass, "Compromise Reached on Navsat," *Aviation Week & Space Technology*, November 26, 1973, 46.

27. Sonnemann, memorandum to Easton.

28. Easton, interview by van Keuren and Tugman.

29. Sonnemann, memorandum to Frazier.

30. Richard Rhodes, *Hedy's Folly: The Life and Breakthrough Inventions of Hedy Lamarr, the Most Beautiful Woman in the World* (New York: Doubleday, 2011), 209.

31. David Holmes, "Navstar Technology," *Countermeasures* 2, no. 12 (December 1976): 53–54.

32. "NavStar Achieves Goals in First Test of Concept," *Aviation Week & Space Technology*, October 17, 1977, 159.

33. Peter Galison, *Einstein's Clocks, Poincare's Maps: Empires of Time* (New York: W. W. Norton, 2003). His evidence has been questioned by Albérto A. Martinez in *Science Secrets: The Truth about Darwin's Finches, Einstein's Wife, and Other Myths* (Pittsburgh: University of Pittsburgh Press, 2011), 206–15. Even if one accepts that Galison overstates the evidence, thought experiments about clock synchronization helped inspire Einstein's formulation of the special theory of relativity.

34. Sonnemann, memorandum to Frazier.

5. Invisible Stars

1. A. D. Richardson, "Making Watches by Machinery," *Harper's New Monthly Magazine* 39 (June–November 1869): 169–82.

2. H. H. Turner, "Greenwich Time," *Cornhill Magazine* 95 (January 1907): 55–69.

3. Parliament.uk, "Living Heritage: Religion and Belief: Witchcraft," http://www.parliament.uk/about/livingheritage/transformingsociety/private-lives/religion/overview/witchcraft (accessed July 5, 2012).

4. Frederick Knight Hunt, "The Planet Watchers of Greenwich," *Harper's New Monthly Magazine* 1 (June–November 1850): 233–37 (reprinted from *Household Words*).

5. Hunt, "Planet Watchers of Greenwich."

6. Philip Edwards, ed., *James Cook: The Journals* (New York: Penguin Classics, 2003), 422.

7. J. C. Beaglehole, *The Life of Captain James Cook* (Stanford CA: Stanford University Press, 1974), 364.

8. U.S. Air Force, "Global Positioning System Fact Sheet," September 15, 2010, http://www.af.mil/information/factsheets/factsheet.asp?id=119.

9. U.S. Air Force, "Global Positioning System Fact Sheet."

10. U.S. Coast Guard Navigation Center, "GPS Constellation Status for 03/28/2013," March 28, 2013, http://www.navcen.uscg.gov/?Do=constellationStatus; Boeing, "Boeing Satellites: Chronology of Launches," April 27, 2012, http://www.boeing.com/defensespace/space/bss/launch/launched2.html.

11. "GPS to Reach Historic High of 30 Operational Spacecraft," *Aerospace Daily and Defense Report*, November 9, 2004, 5, LexisNexis Academic.

12. Garmin, "The GPS Satellite System," http://www8.garmin.com/aboutGPS (accessed July 10, 2012).

13. SatTrackCam Leiden, "A Flashing GPS Satellite (Navstar 39, USA 128)," June 27, 2012, http://sattrackcam.blogspot.com/2012/06/flashing-gps-satellite-navstar-39-usa.html.

14. NASA.gov, "J-Track 3D Satellite Tracking," http://science.nasa.gov/realtime/jtrack/3d/JTrack3D.html (accessed July 10, 2012).

15. GPS.gov, "Space Segment," June 5, 2012, http://www.gps.gov/systems/gps/space.

16. "OCS Legacy History," *GPS World*, June 2008, 28.

17. U.S. Coast Guard Navigation Center, "ICD-GPS-870, Revision A," 6–7, www.navcen.uscg.gov/pdf/gps/ICD_GPS_870_IRN001.pdf (accessed July 10, 2012).

18. GPS.gov, "Control Segment," June 5, 2012, http://www.gps.gov/ systems/gps/control.

19. Craig Covault, "Calling All Global Positioning Sats," *Aviation Week & Space Technology*, October 3, 2005, 30, LexisNexis Academic; GPS.gov, "Control Segment."

20. Covault, "Calling All Global Positioning Sats."

21. Don Branum, "50th Space Wing Completes Transition to New GPS Control System," *Air Force Print News*, September 19, 2007, http://www.afspc .af.mil/news/story.asp?id=123068750; GPS.gov, "Control Segment."

22. IEEE Global History Network, "Heinrich Hertz (1857–1894)," http:// www.ieeeghn.org/wiki/index.php/Heinrich_Hertz_%281857-1894%29 (accessed July 15, 2012).

23. DonRathJr.com, "Audible Range of Human Hearing—Acoustics of Music—Part 2," http://donrathjr.com/audible-range-human-hearing/ (accessed July 15, 2012).

24. DonRathJr.com, "Audible Range of Human Hearing."

25. Federal Communications Commission, "FM Radio," http://www.fcc .gov/topic/fm-radio (accessed July 15, 2012).

26. Department of Defense, *Global Positioning System Standard Positioning Service Performance Standard*, 4th ed. (Washington DC, September 2008), 4, http://www.gps.gov/technical/ps/2008-SPS-performance -standard.pdf.

27. Robert C. Dixon, *Spread Spectrum Systems with Commercial Applications*, 3rd. ed. (New York: John Wiley & Sons, 1994), 439.

28. Rob Walters, *Spread Spectrum: Hedy Lamarr and the Mobile Phone* (Charleston SC: Book Surge, 2005), 241–42.

29. Walters, *Spread Spectrum*, 158.

30. Department of Defense, *Global Positioning System Standard Positioning Service*, 4.

31. James Bao-Yen Tsui, *Fundamentals of Global Positioning System Receivers: A Software Approach* (New York: John Wiley & Sons, 2000), 77.

32. GPS.gov, "GPS Accuracy," http://www.gps.gov/systems/gps/ performance/accuracy/ (accessed July 17, 2012).

33. James Gleick, *Faster: The Acceleration of Just about Everything* (New York: Pantheon Books, 1999), 7.

34. Margaret Coel, "Keeping Time by the Atom," *Invention & Technology* (Winter 1998): 43–48.

35. Michael A. Lombardi, Thomas P. Heavner, and Steven R. Jefferts, "NIST Primary Frequency Standards and the Realization of the SI Second," *Journal of Measurement Science* 2, no. 4 (December 2007): 74–89, http://tf.nist.gov/general/pdf/2039.pdf; Coel, "Keeping Time by the Atom."

36. Coel, "Keeping Time by the Atom."

37. Lombardi, Heavner, and Jefferts, "NIST Primary Frequency Standards"; Coel, "Keeping Time by the Atom."

38. Lombardi, Heavner, and Jefferts, "NIST Primary Frequency Standards."

39. Lombardi, Heavner, and Jefferts, "NIST Primary Frequency Standards."

40. William L. Laurence, "Cosmic Pendulum for Clock Planned," *New York Times*, January 21, 1945, http://select.nytimes.com/gst/abstract.html?res=F20614F8385E1B7A93C3AB178AD85F418485F9.

41. James Jespersen and Jane Fitz-Randolph, *From Sundials to Atomic Clocks: Understanding Time and Frequency* (Washington DC: National Institute of Standards and Technology, March 1999), http://www.boulder.nist.gov/timefreq/general/pdf/1796.pdf; Lombardi, Heavner, and Jefferts, "NIST Primary Frequency Standards"; Coel, "Keeping Time by the Atom."

42. Coel, "Keeping Time by the Atom."

43. Lombardi, Heavner, and Jefferts, "NIST Primary Frequency Standards."

44. Coel, "Keeping Time by the Atom."

45. Coel, "Keeping Time by the Atom"; Lombardi, Heavner, and Jefferts, "NIST Primary Frequency Standards." The General Conference on Weight and Measures defined the second as "the duration of 9,192,631,770 periods of the radiation corresponding to the transition between the two hyperfine levels of the ground state of the cesium-133 atom." The astronomical second, established in 1820, was defined as one "86,400th part of the mean solar day."

46. U.S. Coast Guard Navigation Center, "GPS Constellation Status," http://www.navcen.uscg.gov/?Do=constellationStatus (accessed July 18, 2012).

47. Marvin Epstein, Gerald Freed, and John Rajan, "GPS IIR Rubidium Clocks: In-Orbit Performance Aspects," paper presented at the Thirty-Fifth Annual Precise Time and Time Interval (PTTI) Meeting, San Diego CA, December 2003, http://www.pttimeeting.org/archive meetings/2003papers/paper15.pdf.

48. Department of Defense, *Global Positioning System Standard Positioning Service*, 23.

49. Trimble Navigation Ltd., "Code-Phase vs. Carrier Phase GPS," http://www.trimble.com/gps_tutorial/sub_phases.aspx (accessed July 17, 2012).

6. Going Public

1. Murray Sayle, "KE007: A Conspiracy of Circumstance," *New York Review of Books*, April 25, 1985.
2. Murray Sayle, "Closing the File on Flight 007," *New Yorker*, December 13, 1993, 90–101.
3. Sayle, "KE007."
4. Asaf Degani, "The Crash of Korean Airlines Flight 007," in *Taming HAL: Designing Interfaces beyond 2001* (New York: Palgrave MacMillan, 2003), 60–62.
5. "U.S. Response to Soviet Destruction of KAL Airliner," National Security Decision Directive 102, September 5, 1983, http://www.fas.org/irp/offdocs/nsdd/nsdd-102.htm, also available in the Ronald Reagan Presidential Library, Simi Valley CA.
6. Ronald Reagan, "Remarks at the Annual Convention of the National Association of Evangelicals in Orlando, Florida, March 8, 1983," Public Papers of President Ronald W. Reagan, Ronald Reagan Presidential Library, http://www.reagan.utexas.edu/archives/speeches/1983/30883b.htm.
7. Ronald Reagan, "Address to the Nation on Defense and National Security, March 23, 1983," Public Papers of President Ronald W. Reagan, Ronald Reagan Presidential Library, http://www.reagan.utexas.edu/archives/speeches/1983/32383d.htm.
8. Sharon Watkins Lang, U.S. Army Space & Missile Defense Command/Army Forces Strategic Command (SMD/ARSTRAT) Historical Office, "Where Do We Get 'Star Wars?'" *Eagle*, March 2007, http://www.smdc.army.mil/2008/Historical/Eagle/WheredowegetStarWars.pdf.
9. Kaylene Hughes, "The Army's Precision 'Sunday Punch': The Pershing II and the Intermediate-Range Nuclear Forces Treaty," *Army History* (Fall 2009): 6–16; Federation of American Scientists, "Pershing 2," September 23, 2011, http://www.fas.org/nuke/guide/usa/theater/pershing2.htm.
10. National Priorities Project, "Federal Spending," September 23, 2011, http://nationalpriorities.org/en/resources/federal-budget-101/budget-briefs/federal-spending.

11. Alexander Dallin, *Black Box: KAL 007 and the Superpowers* (Berkeley: University of California Press, 1985), 107.

12. Strobe Talbott et al., "Atrocity in the Skies: KAL Flight 007 Shot Down," *Time*, September 12, 1983, http://www.time.com/time/magazine/article/0,9171,926169,00.html; Sayle, "KE007."

13. A September 7, 2011, Google search of the term "Reagan declassified GPS" yielded 319 results.

14. YouTube.com, "Tim O'Reilly on the SXSW 2011 Stage" (an interview with Jason Calacanis, March 11, 2011, at the South by Southwest interactive, music, and film festival), http://www.youtube.com/watch?v=5x VR5yxjzqU&feature=related (accessed August 30, 2011).

15. Foursquare, "What Is Foursquare?" https://foursquare.com/about (accessed September 8, 2011).

16. Rip and Hasik, *Precision Revolution*, 429. The authors neither cite sources nor provide any explanation for their claim to know the contents of this classified material.

17. Ronald Reagan, "Address before the 38th Session of the United Nations General Assembly in New York, New York, September 26, 1983," Public Papers of President Ronald W. Reagan, Ronald Reagan Presidential Library, http://www.reagan.utexas.edu/archives/speeches/1983/92683a.htm.

18. Tony Rennell, "September 26, 1983: The Day the World Almost Died," *Daily Mail*, December 29, 2007, http://www.dailymail.co.uk/news/article-505009/September-26th-1983-The-day-world-died.html.

19. "Statement by Deputy Press Secretary Larry M. Speakes on the Soviet Attack on a Korean Civilian Airliner," September 16, 1983, Public Papers of President Ronald W. Reagan, Ronald Reagan Presidential Library, http://www.reagan.utexas.edu/archives/speeches/1983/91683c.htm.

20. William P. Clements Jr., deputy secretary of defense, "Memorandum for the Secretaries of the Military Departments; Subject: Defense Navigation Satellite Development Program (DNSDP)," April 17, 1973, 2.

21. U.S. Departments of the Army, Navy and Air Force Navigation Satellite Executive Steering Group, "Memorandum for the Members of NAVSEG; Subject: Miscellany From the Past Chairman," November 2, 1970.

22. Sonnemann, memorandum to Frazier.

23. Parkinson, oral history by Michael Geselowitz.

24. Edward Miguel and Gerard Roland, "The Long Run Impact of Bombing Vietnam," *Journal of Development Economics* 96, no. 1 (September 2011): 1–15.

25. John L. McLucas, "The U.S. Space Program since 1961: A Personal Assessment," in *The U.S. Air Force in Space: 1945 to the 21st Century; Proceedings of the Air Force Historical Foundation Symposium, Andrews AFB, Maryland, September 21–22, 1995,* ed. R. Cargill Hall and Jacob Neufeld (Washington DC: USAF History and Museums Program, United States Air Force), 93.

26. U.S. Energy Information Administration, "Petroleum Chronology of Events 1970–2000, Arab Oil Embargo," May 2002, http://www.eia.gov/pub/oil_gas/petroleum/analysis_publications/chronology/petroleum chronology2000.htm (accessed October 8, 2011); U.S. Department of Labor, Bureau of Labor Statistics, Consumer Price Index, All Urban Consumers (CPI-U), September 15, 2011, ftp://ftp.bls.gov/pub/special .requests/cpi/cpiai.txt; FedPrimeRate.com, "History of the U.S. (Fed) Prime Rate from 1947 to the Present," http://www.wsjprimerate.us/wall_street_journal_prime_rate_history.htm (accessed October 9, 2011).

27. Dana J. Johnson, *Overcoming Challenges to Transformational Space Programs: The Global Positioning System (GPS)* (Arlington VA: Northrop Grumman Analysis Center, 2006), 10, http://www.northropgrumman.com/analysis-center/paper/assets/Overcoming-Challenges-to-Trans.pdf.

28. U.S. General Accounting Office, *Comptroller General's Report to Congress: The NAVSTAR Global Positioning System—A Program with Many Uncertainties* (Washington DC, January 17, 1979), 13–15, http://www .gao.gov/products/PSAD-79-16.

29. Parkinson and Powers, "Fighting to Survive," 8–18.

30. Jeffrey A. Drezner and Giles K. Smith, *An Analysis of Weapon System Acquisition Schedules,* R-3937-ACQ (Santa Monica CA: RAND, 1990), 188.

31. Johnson, *Overcoming Challenges to Transformational Space Programs,* 10.

32. Drezner and Smith, *Analysis of Weapon System Acquisition Schedules,* 188.

33. Matthew E. Skeen, "The Global Positioning System: A Case Study in the Challenges of Transformation," *Joint Force Quarterly,* no. 51 (Fourth Quarter 2008): 88–93. Charles H. Wilson should not be confused with the Texas representative made famous by the movie *Charlie Wilson's War* for funneling arms to Afghanistan's mujahideen during the Soviet occupation.

34. Skeen, "Global Positioning System."

35. Johnson, *Overcoming Challenges to Transformational Space Programs*, 10.

36. Congressional Budget Office, *Strategic Command, Control, and Communications: Alternative Approaches for Modernization* (Washington DC: U.S. Government Printing Office, 1981), 24. Sources vary in spelling out the acronym IONDS; some use the word *detonation*, while others use the word *detection*. CBO used both.

37. Curtis Peebles, *High Frontier: The U.S. Air Force and Military Space Program* (Darby PA: Diane Publishing, 1997), 41–42.

38. Peebles, *High Frontier*, 42.

39. Rockwell International, *GPS/IGS Design Analysis Report*, vol. 1 (Cedar Rapids IA, November 15, 1982), 1. Nuclear detonation detection uses a separate radio signal, L3, at 1,381.05 megahertz.

40. Congressional Budget Office, *Strategic Command, Control, and Communications*, 24.

41. Department of Defense, *Annual Report, Fiscal Year 1981* (Washington DC: U.S. Government Printing Office, 1980), 144.

42. U.S. General Accounting Office, *Comptroller General's Report to Congress: NAVSTAR Should Improve the Effectiveness of Military Missions— Cost Has Increased* (Washington DC, February 15, 1980), 19.

43. Scott Pace et al. *The Global Positioning System: Assessing National Policies* (Santa Monica CA: RAND, 1995), 249; Smithsonian National Museum of American History, "Macrometer v-1000," http://american history.si.edu/collections/surveying/object.cfm?recordnumber =997498 (accessed November 2, 2011).

44. Pace et al., *Global Positioning System*, 248.

45. Richard B. Langley, "Smaller and Smaller: The Evolution of the GPS Receiver," *GPS World*, April 2000, 54–58.

46. Alison Brown, "A Perspective on Land Navigation: The Evolution from Man-Packs to Modules," paper presented at GNSS 2000, Edinburgh, Scotland, May 2000, 1; http://www.navsys.com/Papers/0005003.pdf.

47. Ilari Koskelo, "GNSS: From Experts to Everybody," presented at the European Meeting of the Civil GPS Service Interface Committee, International Information Subcommittee, Stockholm, Sweden, October 27, 2009, http://www.navcen.uscg.gov/pdf/cgsicMeetings/EISub committee/Stockholm_2009/4_Koskelo.pdf.

48. Brown, "Perspective on Land Navigation," 1.

49. Langley, "Smaller and Smaller."

50. Sheryl Mikkola, "Generalized Development Model (GDM)," Institute of Navigation, Navigation Museum, http://www.ion.org/museum/cat _view.cfm?cid=7&scid=9 (accessed January 5, 2011).

51. Phil Ward, "Texas Instruments TI 4100 NAVSTAR Navigator," Institute of Navigation, Navigation Museum, http://www.ion.org/museum/ item_view.cfm?cid=7&scid=9&iid=22 (accessed January 5, 2011).

52. "Texas Instruments Introduces New Satplan Software for Global Positioning System," *Maritime Reporter and Engineering News*, March 1, 1985, 40.

53. Old-Computers.com, "Texas Instruments Portable Professional Computer," http://www.old-computers.com/museum/computer.asp?st =1&c=472 (accessed January 8, 2012).

54. U.S. General Accounting Office, *Comptroller General's Report to Congress: NAVSTAR Should Improve the Effectiveness of Military Missions*, 12.

55. NASA.gov, "Space Shuttle Missions," http://www.nasa.gov/mission _pages/shuttle/shuttlemissions/index.html (accessed August 29, 2011).

56. Frank Czopek, "GPS 12," Institute of Navigation, Navigation Museum, http://www.ion.org/museum/item_view.cfm?cid=7&scid=9&iid=23 (accessed January 8, 2011).

57. Pace et al., *Global Positioning System*, 243; Mark C. Cleary, *Delta II & III Space Operations at Cape Canaveral, 1989–2009* (Patrick AFB FL: Forty-Fifth Space Wing Office of History, 2009), 4, www.afspacemuseum .org/archive/histories/Delta.pdf.

58. Cleary, *Delta II & III Space Operations*, 2, 4.

59. GPS.gov, "Space Segment," http://www.gps.gov/systems/gps/space (accessed August 29, 2011).

60. Pace et al., *Global Positioning System*, 249.

61. Langley, "Smaller and Smaller."

62. Langley, "Smaller and Smaller."

63. Langley, "Smaller and Smaller."

7. Going to War

1. John A. Tirpak, "The Secret Squirrels," *Air Force Magazine* 77, no. 5 (April 1994): 56–60.

2. Rip and Hasik, *Precision Revolution*, 160.

3. Jon Lake, B-52 *Stratofortress Units in Operation Desert Storm* (Oxford UK: Osprey, 2004), 35.

4. Joseph C. Jones, "Surprised by Secret Squirrel," *Air Force Print News Today*, February 1, 2011, http://www.307bw.afrc.af.mil/news/story _print.asp?id=123240704.

5. Tirpak, "Secret Squirrels."

6. Rip and Hasik, *Precision Revolution*, 159.

7. George M. Siouris, *Missile Guidance and Control Systems* (New York: Springer-Verlag, 2003), 551–52.

8. John T. Nielson, "The Untold Story of the CALCM: The Secret GPS Weapon Used in the Gulf War," *GPS World* 6, no. 1 (January 1995): 26–32.

9. John T. Nielson, "CALCM—The Untold Story of the Weapon Used to Start the Gulf War," *IEEE Aerospace and Electronic Systems Magazine* 9, no. 7 (July 1994): 18–22.

10. U.S. Air Force, "AGM-B/C/D Missiles Factsheet," May 24, 2010, http:// www.af.mil/information/factsheets/factsheet.asp?id=74.

11. Tirpak, "Secret Squirrels."

12. Richard Hallion, *Storm over Iraq: Air Power and the Gulf War* (Washington DC: Smithsonian Institution Press, 1992), 102–3; Congressional Budget Office, *Naval Combat Aircraft: Issues and Options* (Washington DC: Government Printing Office, November 1987), 11.

13. Rip and Hasik, *Precision Revolution*, 167.

14. Hallion, *Storm over Iraq,* 296; Rip and Hasik, *Precision Revolution,* 168; MissileThreat.com, "Cruise Missiles: AGM-84E SLAM," February 13, 2012, http://www.missilethreat.com/cruise/id.119/cruise_detail.asp.

15. U.S. Navy, "Fact File: SLAM-ER Missile," February 20, 2009, http:// www.navy.mil/navydata/fact_display.asp?cid=2200&tid=1100&ct=2.

16. Hallion, *Storm over Iraq,* 296; Rip and Hasik, *Precision Revolution,* 168.

17. U.S. Department of Defense, *Conduct of the Persian Gulf War, Final Report to Congress* (Washington DC: U.S. Government Printing Office, 1992), 858–60; Rip and Hasik, *Precision Revolution,* 167–70.

18. Tirpak, "Secret Squirrel"; Nielson, "Untold Story of the CALCM"; Rip and Hasik, *Precision Revolution,* 159; Hallion, *Storm over Iraq,* 172.

19. "Air Force Launched 35 ALCMs on First Night of Gulf Air War," *Defense Daily,* January 17, 1992.

20. U.S. Department of Defense, *Conduct of the Persian Gulf War*, 853.

21. *Gulf War Air Power Survey*, vol. 4, *Weapons, Tactics, and Training* (Washington DC: U.S. Government Printing Office, 1993), 248–52.

22. U.S. Air Force, "AGM-B/C/D Missiles Factsheet."

23. U.S. Department of Defense, *Conduct of the Persian Gulf War*, 215; Naval Air Systems Command, "Aircraft & Weapons: Tomahawk," http://www.navair.navy.mil/index.cfm?fuseaction=home.display&key =F4E98B0F-33F5-413B-9FAE-8B8F7C5F0766 (accessed February 16, 2012).

24. U.S. Department of Defense, *Conduct of the Persian Gulf War*, 171, 863.

25. "U.S. Military Technology in Saudi Arabia," *CBS Evening News*, transcript, November 16, 1990, LexisNexis Academic.

26. "Soviet Crackdown Leaves 13 Dead in Lithuania," *ABC Weekend News Sunday*, transcript, January 13, 1991, LexisNexis Academic.

27. "Spy Satellites Have Role in Gulf War," *CBS Evening News*, transcript, February 8, 1991, LexisNexis Academic.

28. Karen Tumulty and Bob Drogin, "Ground War Puts Some Exotic New Weapons Systems to Test," *Los Angeles Times*, February 25, 1991, http://articles.latimes.com/print/1991-02-25/news/mn-1481_1_ground-war.

29. Andrew Pollack, "Business Technology: War Spurs Navigation by Satellite," *New York Times*, February 6, 1991.

30. *Crossfire*, CNN, transcript, January 25, 1991, LexisNexis Academic.

31. "On Fighting a High Tech Ground War," *CBS News Special Report*, transcript, February 23, 1991, LexisNexis Academic.

32. Thomas B. Allen, F. Clifton Berry, and Norman Polmar, *CNN: War in the Gulf* (Atlanta: Turner, 1991), 146.

33. Allen, Berry, and Polmar, *CNN*, 180.

34. Rick Atkinson, *Crusade: The Untold Story of the Persian Gulf War* (New York: Houghton Mifflin, 1993), 76.

35. *Gulf War Air Power Survey*, 4:85.

36. U.S. General Accounting Office, *Operation Desert Storm: Evaluation of the Air Campaign* (Washington DC, June 1997), 178, http://www.gao .gov/products/NSIAD-97-134.

37. Kenneth P. Werrell, *Chasing the Silver Bullet: U.S. Air Force Weapons Development from Vietnam to Desert Storm* (Washington DC: Smithsonian Books, 2003), 262.

38. Werrell, *Chasing the Silver Bullet*, 264; U.S. Department of Defense, *Con-*

duct of the Persian Gulf War, 785; *Gulf War Air Power Survey, Summary Report* (Washington DC: U.S. Government Printing Office, 1993), 199.

39. Walter J. Boyne, *Beyond the Wild Blue: A History of the U.S. Air Force 1947–1997* (New York: St. Martin's Press, 1997), 275.

40. Darrell D. Whitcomb, *Combat Search and Rescue in Desert Storm* (Maxwell AFB AL: Air University Press, 2006), 113.

41. *Gulf War Air Power Survey, Summary Report*, 199; *Gulf War Air Power Survey*, vol. 2, *Operations and Effects and Effectiveness* (Washington DC: U.S. Government Printing Office, 1993), 236.

42. Richard Mackenzie, "Apache Attack," *Air Force Magazine*, October 1991, http://www.airforce-magazine.com/MagazineArchive/Pages/1991/October%201991/1091apache.aspx.

43. Thomas Taylor, *Lightning in the Storm: The 101st Air Assault Division in the Gulf War* (New York: Hippocrene Books, 1994), 158, 169.

44. Taylor, *Lightning in the Storm*, 132.

45. Hallion, *Storm over Iraq*, 167.

46. Whitcomb, *Combat Search and Rescue*, 202–7.

47. Whitcomb, *Combat Search and Rescue*, 207.

48. Edward J. Marolda and Robert John Schneller Jr., *Shield and Sword: The United States Navy and the Persian Gulf War* (Annapolis: Naval Institute Press, 2001), 301.

49. Marolda and Schneller, *Shield and Sword*, 247.

50. Rip and Hasik, *Precision Revolution*, 419.

51. U.S. Department of Defense, *Conduct of the Persian Gulf War*, 336; Rip and Hasik, *Precision Revolution*, 178–79.

52. Maryann Lawlor, "Keeping Track of the Blue Force," *Signal Magazine*, July 2003, http://www.afcea.org/content/?q=node/127; Roxana Tiron, "Army's Blue-Force Tracking Technology Was a Tough Sell," *National Defense and Technology Magazine*, December 2003, http://www.nationaldefensemagazine.org/archive/2003/December/Pages/Armys_Blue 3685.aspx; StrategyPage.com, "Grandson of Blue Force Tracker Goes to War," February 28, 2013, http://www.strategypage.com/htmw/htiw/articles/20130228.aspx.

53. H. Norman Schwarzkopf and Peter Petre, *It Doesn't Take a Hero* (New York: Bantam Books, 1993), 455.

54. U.S. Department of Defense, *Conduct of the Persian Gulf War*, 219.

55. U.S. Department of Defense, *Conduct of the Persian Gulf War*, 219.

56. Roger Cohen and Claudio Gatti, *In the Eye of the Storm: The Life of General H. Norman Schwarzkopf* (New York: Farrar, Straus, and Giroux, 1991), 289.

57. Kaleb Dissinger, "GPS Goes to War: The Global Positioning System in Operation Desert Storm," Army.mil, February 14, 2008, http://www .army.mil/article/7457/GPS_Goes_to_War__The_Global_Positioning _System_in_Operation_Desert_Storm_/.

58. U.S. Department of Defense, *Conduct of the Persian Gulf War,* 654.

59. Dissinger, "GPS Goes to War."

60. Sheryl Mikkola, "Rockwell Manpack Global Positioning System GPS Receiver," Institute of Navigation, Navigation Museum, http://www.ion .org/museum/item_view.cfm?cid=7&scid=9&iid=10 (accessed March 13, 2012); Smithsonian National Museum of American History, "Manpack Global Positioning System GPS Receiver," http://americanhistory.si.edu/ collections/object.cfm?key=35&objkey=220 (accessed March 13, 2012).

61. Rip and Hasik, *Precision Revolution,* 136.

62. U.S. Space Command, *United States Space Command: Operations Desert Shield and Desert Storm Assessment* (Peterson AFB CO, January 1992), 28.

63. Dissinger, "GPS Goes to War."

64. U.S. Department of Defense, *Conduct of the Persian Gulf War,* 876.

65. *Gulf War Air Power Survey,* vol. 5, *A Statistical Compendium and Chronology* (Washington DC: U.S. Government Printing Office, 1993), 132.

66. *Gulf War Air Power Survey,* 5:131; Whitcomb, *Combat Search and Rescue,* 245.

67. Dissinger, "GPS Goes to War."

68. Rip and Hasik, *Precision Revolution,* 138.

69. Rip and Hasik, *Precision Revolution,* 147.

70. U.S. Space Command, *United States Space Command: Operations Desert Shield and Desert Storm Assessment,* 40.

71. Rip and Hasik, *Precision Revolution,* 137–42.

72. GenCorp Aerojet Space Systems Division, *DSP Desert Storm Summary Briefing* (Azusa CA, June 1991), 32–35.

73. Craig Covault et al., "Desert Storm Reinforces Military Space Directions," *Aviation Week & Space Technology* 134, no. 14 (April 8, 1991): 42; U.S. Department of Defense, *Report of the Secretary of Defense to the President and the Congress* (Washington DC: U.S. Government Printing Office, February 1992), 85.

74. U.S. Space Command, *United States Space Command: Operations Desert Shield and Desert Storm Assessment*, 25–26.

75. Rip and Hasik, *Precision Revolution*, 132.

76. U.S. Space Command, *United States Space Command: Operations Desert Shield and Desert Storm Assessment*, 27; Rip and Hasik, *Precision Revolution*, 132.

77. U.S. Space Command, *United States Space Command: Operations Desert Shield and Desert Storm Assessment*, 28.

78. A complete explanation of the command structure and history of the 2nd Space Operations Squadron is available at U.S. Air Force, Schriever AFB, "2nd Space Operations Squadron," February 8, 2013, http://www.schriever.af.mil/library/factsheets/factsheet.asp?id=4045.

79. Rip and Hasik, *Precision Revolution*, 133–34.

80. U.S. Space Command, *United States Space Command: Operations Desert Shield and Desert Storm Assessment*, 27.

81. Rip and Hasik, *Precision Revolution*, 133–34.

82. Rip and Hasik, *Precision Revolution*, 133–34.

83. U.S. Department of Defense, *Conduct of the Persian Gulf War*, 877.

84. Johnson, *Overcoming Challenges to Transformational Space Programs*, 10.

85. Edwin Chen, "Capital Mood Mixes Anger, Resignation," *Los Angeles Times*, January 17, 1991, http://articles.latimes.com/1991-01-17/news/mn-428_1_national-capital.

86. Rip and Hasik, *Precision Revolution*, 224.

87. General Dynamics Ordnance and Tactical Systems, "General Dynamics and BAE Systems Demonstrate 81mm Precision Mortar Round," news release, February 21, 2012, http://www.gd-ots.com/News.html.

8. Going Mainstream

1. Except where noted otherwise, this account of the shoot down and rescue of Capt. Scott O'Grady is drawn from Scott O'Grady, *Return with Honor* (New York: Doubleday, 1995); and Mary Pat Kelly, *Good to Go: The Rescue of Scott O'Grady from Bosnia* (Annapolis MD: Naval Institute Press, 1996).

2. Dan Stets, "Where on Earth? The Global Positioning System Can Answer That Question," *Philadelphia Inquirer*, May 16, 1996, Lexis-Nexis Academic.

3. Debra Gersh Hernandez, "Public Interest Waning in O.J. Trial: Times Mirror Center Survey Shows Number of People Following It Very Closely Is at the Lowest Point since June 1994," *Editor & Publisher*, July 8, 1995, 12, LexisNexis Academic.

4. Haya El Nasser, "High-Flying O'Grady Fills Hunger for Hero," *USA Today*, June 14, 1995, LexisNexis Academic.

5. Associated Press, "1995: A Year of Trials and Tribulations; Top 10 Lists," *Austin-American Statesman* (TX), December 31, 1995, LexisNexis Academic. O'Grady's rescue ranked eighth. The bombing of the Alfred P. Murrah Federal Building in Oklahoma City was the top story. O. J. Simpson's acquittal ranked third, after the breakup of Yugoslavia.

6. Rowan Scarborough, "Downed Pilot's Radio Called Obsolete: Air Force Told State-of-the-Art Model Is Available," *Washington Times*, July 17, 1995, LexisNexis Academic.

7. Scarborough, "Downed Pilot's Radio Called Obsolete."

8. Scarborough, "Downed Pilot's Radio Called Obsolete."

9. "Add Satcom to Rescue Radios—Fast," editorial, *Aviation Week & Space Technology*, October 23, 1995, 78, LexisNexis Academic.

10. "Add Satcom to Rescue Radios."

11. "Rockwell Wins Combat Survival Radio Project," *Aerospace Daily*, February 29, 1996, 316, LexisNexis Academic.

12. "GPS Companies Awarded Combat Survivor Evader Locator Contracts," *Global Positioning & Navigation News*, March 7, 1996, LexisNexis Academic.

13. David Atkinson, "DoD Report Says Current CSEL 'Not Operationally Suitable,'" *Defense Daily*, February 11, 1999, LexisNexis Academic.

14. "Critical Intelligence," *Inside the Pentagon*, October 7, 2004, LexisNexis Academic.

15. "Program Office Marks Upcoming Delivery of 50,000th Combat Survivor Evader Locator Unit," *U.S. Fed News*, October 5, 2011, LexisNexis Academic.

16. Ann Roosevelt, "Personnel Recovery Goes Beyond Altruism, General Says," *Defense Daily*, September 1, 2004, LexisNexis Academic.

17. "Rockwell Wins $13 Million Contract for Pilot Location System," *Armed Forces Newswire Service*, March 5, 1996, LexisNexis Academic; "Program Office Marks Upcoming Delivery."

18. Elizabeth Rees, "Due to 'Urgent' Need, USAF to Produce More Survi-

vor Radios in FY-03," *Inside the Air Force*, April 25, 2003, LexisNexis Academic.

19. James R. Schlesinger et al., *The Global Positioning System: Charting the Future, Summary Report* (Washington DC: National Academy of Public Administration and National Research Council, May 1995), http://www.navcen.uscg.gov/?pageName=gpsFuture.

20. Pace et al., *Global Positioning System*.

21. Bruce D. Nordwall, "Navsat Users Want Civil Control," *Aviation Week & Space Technology*, October 18, 1993, 57, LexisNexis Academic.

22. Anne G. K. Solomon, "The Global Positioning System," in *Triumphs and Tragedies of the Modern Presidency: Seventy-Six Case Studies in Presidential Leadership*, ed. David M. Abshire (Westport CT: Praeger, 2001), 109.

23. Patricia A. Gilmartin, "Gulf War Rekindles U. S. Debate on Protecting Space System Data," *Aviation Week & Space Technology*, April 29, 1991, 55, LexisNexis Academic.

24. Ben Iannotta, "Europe Weighs Building Its Own GPS," *Aerospace America*, September 1, 1999, 34, LexisNexis Academic.

25. Jeffrey M. Lenorovitz, "Steady Growth Seen for Commercial Space," *Aviation Week & Space Technology*, March 15, 1993, 83, LexisNexis Academic.

26. Laurence Hooper, Jacob M. Schlesinger, and Richard L. Hudson, "Precise Navigation Points to New Worlds," *Wall Street Journal*, March 4, 1991, ProQuest.

27. Hooper, Schlesinger, and Hudson, "Precise Navigation Points to New Worlds."

28. Hooper, Schlesinger, and Hudson, "Precise Navigation Points to New Worlds."

29. John T. Mulqueen, "Clocking the Network—Synchronization Needs Expanding," *InternetWeek*, November 11, 1991, 23, LexisNexis Academic.

30. Gilmartin, "Gulf War Rekindles U. S. Debate"; John J. Fialka, "Poor Man's Cruise: Airliners Can Exploit U.S. Guidance System, but So Can Enemies," *Wall Street Journal*, August 26, 1993, ProQuest.

31. Pace et al., *Global Positioning System*, 150.

32. Pace et al., *Global Positioning System*, 14.

33. Fialka, "Poor Man's Cruise."

34. Fialka, "Poor Man's Cruise."

35. Philip J. Klass, "INMARSAT Plan Spurs GPS Debate," *Aviation Week & Space Technology*, July 26, 1993, 23, LexisNexis Academic.

36. Pace et al., *Global Positioning System*, 180.

37. Pace et al., *Global Positioning System*, 289.

38. U.S. Coast Guard Navigation Center, "GPS Fully Operational Statement of 1995," May 2, 2001, http://www.navcen.uscg.gov/?pageName=global.

39. Government Printing Office, "59 FR-Announcement of Global Positioning System (GPS) Initial Operational Capability (IOC) and Its Impact on Vessel Carriage Requirement Regulations," *Federal Register* 59, no. 56 (March 23, 1994), http://www.gpo.gov/fdsys/pkg/FR-1994-03-23/html/94-6813.htm.

40. U.S. Coast Guard Navigation Center, "NDGPS General Information," http://www.navcen.uscg.gov/?pageName=dgpsMain (accessed July 1, 2012).

41. "GPS Approved for IFR Approaches," *Aviation Week & Space Technology*, June 14, 1993, 44, LexisNexis Academic.

42. Bruce D. Nordwall, "FAA Details Phases for GPS Flight Use," *Aviation Week & Space Technology*, June 21, 1993, 75, LexisNexis Academic.

43. Federal Aviation Administration, "Global Positioning System: A Guide for the Approval of GPS Receiver Installation and Operation," October 1996, 312, http://www.faa.gov/about/office_org/headquarters_offices/avs/offices/afs/afs400/afs470/gps_guide/media/gps1.pdf; David Hughes, "FAA Gears Up for GPS Approaches," *Aviation Week & Space Technology*, September 19, 1994, 65, LexisNexis Academic.

44. Federal Aviation Administration, "Global Positioning System."

45. Garmin, "The Garmin 430: An Introduction to Aviation GPS: Receiver Autonomous Integrity Monitoring," May 25, 2012, http://www.garmin430.com/raim.htm.

46. Daniel Pearl, "FAA Clears Civilian Airlines to Use Military Satellite Signals in Navigation," *Wall Street Journal*, February 18, 1994, ProQuest.

47. Nordwall, "FAA Details Phases for GPS Flight Use."

48. David Hughes, "Continental Pursues GPS-Only Approaches," *Aviation Week & Space Technology*, August 2, 1993, 57, LexisNexis Academic.

49. Hughes, "Continental Pursues GPS-Only Approaches."

50. Nordwall, "FAA Details Phases for GPS Flight Use."

51. AOPA Online, "Airports and Landing Areas, 1965–2009," March 25, 2011, http://www.aopa.org/whatsnew/stats/airports.html.

52. AOPA Online, "Airports and Landing Areas, 1965–2009."

53. Bureau of Transportation Statistics, "Historical Air Traffic Statistics, Annual 1954–1980," May 30, 2012, http://www.bts.gov/programs/airline_information/air_carrier_traffic_statistics/airtraffic/annual/1954_1980.html; Bureau of Transportation Statistics, "Historical Air Traffic Statistics, 1981–1995," May 30, 2012, http://www.bts.gov/programs/airline_information/air_carrier_traffic_statistics/airtraffic/annual/1981_present.html.

54. Bureau of Transportation Statistics, "Historical Air Traffic Statistics, 1981–1995"; Andrew R. Thomas, *Soft Landing: Airline Industry Strategy, Service, and Safety* (New York: Apress Media, 2011), 54–55.

55. Thomas, *Soft Landing*, 77; U.S. Government Accounting Office, "Commuter Airline Safety Would Be Enhanced with Better FAA Oversight," March 17, 1992, http://www.gao.gov/products/T-RCED-92-40.

56. Thomas, *Soft Landing*, 55.

57. Federal Document Clearing House, Congressional Testimony, "Testimony March 24, 1994: Jeff Ariens, Director of Flight Operations Technology, Continental Airlines, House Science, Global Positioning System," LexisNexis Academic.

58. Federal Document Clearing House, Ariens testimony.

59. Federal Document Clearing House, Ariens testimony.

60. Federal Document Clearing House, Congressional Testimony, "Testimony March 24, 1994: Martin T. Pozesky, Associate Administrator, System Engineering and Development, House Science, Global Positioning System," LexisNexis Academic.

61. Federal Aviation Administration, "Advisory Circular 120-29A: Criteria for Approval of Category I and Category II Weather Minima for Approach," August 17, 2010, http://www.faa.gov/regulations_policies/advisory_circulars/index.cfm/go/document.information/documentID/22752.

62. Federal Document Clearing House, Pozesky testimony.

63. Nordwall, "FAA Details Phases for GPS Flight Use."

64. Hughes, "FAA Gears Up for GPS Approaches."

65. Daniel Pearl, "FAA May Spend Up to $500 Million to Track Aircraft," *Wall Street Journal*, June 9, 1994, ProQuest.

66. "President Clinton Says U.S. Committed to Civil Use of GPS," *Aviation Daily*, March 28, 1995, LexisNexis Academic.

67. Schlesinger et al., *Global Positioning System*, 7.

68. Pace et al., *Global Positioning System*, 117.

69. Pace et al., *Global Positioning System*, 211.

70. Pace et al., *Global Positioning System*, 104.

71. Schlesinger et al., *Global Positioning System*, 9.

72. Federation of American Scientists, "Presidential Decision Directive NTSC 6," March 26, 1996, http://www.fas.org/spp/military/docops/national/gps.htm (accessed June 3, 2012).

73. "Press Conference with Vice President Al Gore, Transportation Secretary Federico Pena and Others, Subject: Global Positioning System, The Old Executive Office Building," *Federal News Service*, March 29, 1996, LexisNexis Academic.

74. Matthew L. Wald, "A Contract Is Awarded to Improve Navigation," *New York Times*, August 4, 1995.

75. Bruce D. Nordwall, "FAA Swaps WAAS Vendors," *Aviation Week & Space Technology*, May 6, 1996, 34, LexisNexis Academic; PR Newswire, "AOPA Applauds FAA Decisiveness in Cancelling Wilcox WAAS Contract but Criticizes Slow Implementation of GPS Satellite Navigation," news release, April 26, 1996, LexisNexis Academic.

76. Perry Bradley, "Wide Area Augmentation," *Business & Commercial Aviation*, November 1997, 86, LexisNexis Academic.

77. James E. Swikard, ed., "The FAA Officially Turned on the Wide Area Augmentation System," *Business & Commercial Aviation*, August 2003, LexisNexis Academic; Matthew L. Wald, "The Long Search for the Perfect Landing," *International Herald Tribune*, June 11, 2003, LexisNexis Academic.

78. "Wide Area Augmentation System Instrument Approaches Now Outnumber Instrument Landing System Approaches," *U.S. Fed News*, November 6, 2008, LexisNexis Academic.

79. U.S. Coast Guard Navigation Center, "Nationwide Differential Global Positioning (NDGPS) Memorandum of Agreement," http://www.navcen.uscg.gov/pdf/ndgps/ndgpsESC/backgroundDocuments/NDGPS%20MOA%20%20Feb%2099.pdf (accessed July 1, 2012).

80. "WAAS Receivers Sell Well—Is Coast Guard System Necessary," *Global Positioning & Navigation News*, September 6, 2000, LexisNexis Academic.

81. Timothy A. Klein, "Nationwide Differential GPS (NDGPS) Program Update," PowerPoint presentation at the Fiftieth Meeting of the Civil GPS Service Interface Committee, ION GNSS 2010 Conference, Portland OR, September 21, 2010, http://www.navcen.uscg.gov/pdf/ndgps/50th_CGSIC_NDGPS_Klein_092110.pdf.

82. Pace et al., *Global Positioning System*, 108.

83. Pace et al., *Global Positioning System*, 109.

84. Neela Banerjee, "Tracking Travel," *Wall Street Journal*, January 4, 1994, ProQuest.

85. Business Wire, "Satellite Navigation Leader Magellan Buys Rockwell International Driver Information Systems Group; Targets the $1 Billion Private-Passenger-Vehicle Navigation Market," news release, August 4, 1997, LexisNexis Academic.

86. Business Wire, "Satellite Navigation Leader Magellan Buys Rockwell International Driver Information Systems Group."

87. Business Wire, "Magellan Systems Corp. Forms U.K. Subsidiary, Expands European Marketing and Sales," news release, December 18, 1991, LexisNexis Academic; MiTAC International Corp., "Magellan: About Us," http://www.magellangps.com/About-Us (accessed May 17, 2012).

88. Business Wire, "Magellan Systems Corp. forms U.K. Subsidiary."

89. Trimble Navigation Ltd., "Company History," http://www.trimble.com/corporate/about_history.aspx (accessed May 14, 2012).

90. Business Wire, "Trimble Navigation Limited Completes Initial Public Offering," news release, July 20, 1990, LexisNexis Academic.

91. John T. McQueen, "Life Is a Box of Satellites," *InternetWeek*, August 14, 1995, 54, LexisNexis Academic.

92. Trimble Navigation Ltd., "Company History."

93. Amy Gilroy, "Car Navigation Market Gaining Speed," *TWICE*, January 2004, http://www.twice.com/article/254755-Car_Navigation_Market_Gaining_Speed.php.

94. Gilroy, "Car Navigation Market Gaining Speed."

95. "2003-2004 Market Share Reports by Category," *TWICE*, January 6, 2005, http://www.twice.com/article/244763 2003_2004_Market_Share_Reports_By_Category.php.

96. "2003-2004 Market Share Reports by Category."

97. Amy Gilroy, "Navigation Sales Still Climbing," *TWICE*, January 6, 2005,

http://www.twice.com/article/235087-Navigation_Sales_Still
_Climbing.php.

98. Gilroy, "Navigation Sales Still Climbing."

99. "2005 Market Share Reports by Category," TWICE, January 5, 2006,
http://www.twice.com/article/2557292005_Market_Share_Reports_By
_Category.php.

100. "2005 Market Share Reports by Category."

101. Garmin Ltd., *2000 Annual Report* (Grand Cayman, Cayman Islands,
2001), 6.

102. Garmin Ltd., *2000 Annual Report*, 8.

103. Garmin Ltd., *2000 Annual Report*, 5, 14.

104. Garmin Ltd., *2012 Annual Report* (Schaffhausen, Switzerland, 2012), 11.

105. Garmin Ltd., *2012 Annual Report*, 11.

106. This entire account of TomTom's corporate history is from TomTom,
"History," December 2010, http://corporate.tomtom.com/history.cfm;
and TomTom International BV, *2013 Annual Report* (Amsterdam, 2013),
78–79.

107. Amy Gilroy, "Deluge of Portable GPS Debuts for Fall," TWICE, October
9, 2006, http://www.twice.com/article/248069-Deluge_Of_Portable
_GPS_Debuts_For_Fall.php.

108. Gilroy, "Deluge of Portable GPS Debuts."

109. Doug Olenick, "Holiday Sales Season Saw Slow Steady Growth: NPD,"
TWICE, December 18, 2006, http://www.twice.com/article/250339
Holiday_Sales_Season_Saw_Slow_Steady_Start_NPD.php.

110. Amy Gilroy, "PNDs Get Computer, Cellular Upgrades," TWICE, January
7, 2008, http://www.twice.com/article/257622-PNDs_Get_Computer
_Cellular_Upgrades.php.

111. Gilroy, "PNDs Get Computer, Cellular Upgrades"; Amy Gilroy, "Garmin Tops TomTom in Worldwide PNDs," TWICE, March 19, 2008,
http://www.twice.com/article/262209-Garmin_Tops_TomTom_In
_Worldwide_PNDs.php.

112. Gilroy, "Garmin Tops TomTom in Worldwide PNDs."

113. Walter S. Mossberg, "A New Device Makes Your No.@*! Cell Phone a
Little Bit Smarter," *Wall Street Journal*, June 24, 1999, ProQuest.

114. Dawn C. Chmielewski, "Map Services: Don't Try This While Driving,"
San Jose Mercury News, November 7, 2005, LexisNexis.

115. Amy Gilroy, "Dealers Resent Snub by Apple at iPhone Launch,"

TWICE, July 2, 2007, http://www.twice.com/article/263215Dealers
_Resent_Snub_By_Apple_At_iPhone_Launch.php.

116. Greg Tarr, "iPhone Becomes AT&T's No. 1 Phone," *TWICE*, October 19, 2007, http://www.twice.com/article/245688-iPhone_Becomes_AT_T_s _No_1_Phone.php.

117. Amy Gilroy, "Smartphones Capture Almost 20% of Cellphone Market," *TWICE*, September 8, 2008, http://www.twice.com/article/ 240887Smartphones_Capture_Almost_20_Of_Cellphone_Market.php.

118. Gilroy, "Smartphones Capture Almost 20% of Cellphone Market."

119. Amy Gilroy, "iPhone Attracts Plenty of GPS-Related Apps," *TWICE*, August 25, 2009, http://www.twice.com/article/328762-iPhone_Attracts _Plenty_Of_GPS_Related_Apps.php.

120. Gilroy, "iPhone Attracts Plenty of GPS-Related Apps."

121. Janice Partyka, "Hitting the Mainstream," *GPS World*, May 2009, 18, Business Source Complete, EBSCOhost; Gilroy, "iPhone Attracts Plenty of GPS-Related Apps."

122. Walter S. Mossberg, "Motorola's Droid Is Smart Success for Verizon Users," *Wall Street Journal*, November 5, 2009, ProQuest.

123. Roger Cheng and Niraj Sheth, "Motorola's Droid Makes Its Debut," *Wall Street Journal*, October 29, 2009, ProQuest.

124. Gustav Sandstrom, "Nokia Launches Free Navigation Services," *Wall Street Journal*, January 21, 2010, ProQuest.

125. David Barkholz and Leslie J. Allen, "Rivalry Heats Up in Navigation Device Market," *Automotive News*, February 1, 2010, MasterFILE Complete, EBSCOhost.

126. Business Wire, "PND Vendors Shifting Strategies."

127. Joseph Palenchar, "iSuppli: Smartphone Impact Seen on In-Vehicle Systems," *TWICE*, September 13, 2010, http://www.twice.com/ article/457033-iSuppli_Smartphone_Impact_Seen_On_In_Vehicle _Systems.php.

128. Palenchar, "iSuppli."

129. Archibald Preuschat and Anton Troianovski, "Smartphones Sting TomTom," *Wall Street Journal*, June 28, 2011, ProQuest; Berg Insight, "Mobile Navigation Services and Devices—5th Edition, Executive Summary," http://www.berginsight.com/ReportPDF/Summary/ bi-mns5-sum.pdf (accessed June 10, 2012).

130. Berg Insight, "Mobile Navigation Services and Devices."

131. Clement J. Driscoll, "Locating Wireless 911 Callers Is Goal of Industry Associations," *Radio Communications Report*, September 19, 1994, LexisNexis Academic.

132. Driscoll, "Locating Wireless 911 Callers."

133. Jeffrey Silva, "Public Safety, Wireless Industry at Odds Over Feasibility of E911,"*Radio Communications Report*, January 30, 1995, LexisNexis Academic.

134. "FCC Adopts Rules to Implement Enhanced 911 for Wireless Services," *Communications Today*, June 13, 1996, LexisNexis Academic.

135. "FCC Adopts Rules."

136. "FCC Adopts Rules."

137. Peter Clarke, "Radio Triangulation Proposed for Cellular 911 Call," *Electronic Engineering Times*, March 9, 1998, LexisNexis Academic.

138. Clarke, "Radio Triangulation Proposed."

139. Business Wire, "SiRF Delivers Instant Location Information for Consumer Products; Start-Up to Commercialize GPS Technology for a Wide Range of Consumer Applications," news release, September 10, 1996, LexisNexis Academic.

140. Elizabeth V. Mooney, "Handset Makers Set to Ride GPS Technology Wave with SiRF," *Radio Communications Report*, August 17, 1998, LexisNexis Academic.

141. John Markoff, "Deals to Move Global Positioning Technology toward Everyday Use," *New York Times*, August 10, 1998, http://www.nytimes .com/1998/08/10/business/deals-to-move-global-positioning -technology-toward-everyday-use.html.

142. Markoff, "Deals to Move Global Positioning Technology."

143. Darrell Dunn, "RFMD Takes Aim at GPS—To Buy IBM Unite to Increase Exposure in Cell Phone Market," *Electronic Buyer's News*, December 10, 2001, LexisNexis Academic.

144. Rebecca Sykes, "Qualcomm Buys SnapTrack for $1 Billion," *Network World*, January 26, 2000, LexisNexis Academic; Donna Fuscaldo, "Chip Always in Touch with Cellphones," *Wall Street Journal*, November 20, 2002, ProQuest; David Pringle and Pui-Wing Tam, "Nokia and Qualcomm Square Off—Market Share Battle Intensifies for Dominance of 3G Mobile Phone Technology," *Wall Street Journal*, February 20, 2003, ProQuest.

145. Fuscaldo, "Chip Always in Touch with Cellphones."

146. Linley Gwennap, "Broadcom Gains Ground in the GPS Chip Market," *Electronic Design*, September 9, 2010, 16, Business Source Complete, EBSCOhost.

147. "GPS Chipmakers Headed for More Consolidation?" *GPS World*, July 2008, 27, Business Source Complete, EBSCOhost.

148. Michael V. Copeland, "Location, Location, Location," *Fortune*, November 12, 2007, 147–50, Business Source Complete, EBSCOhost.

149. Jonathan Marino, "CSR Buys SiRF Technology," *Mergers and Acquisitions Report*, February 16, 2009, 3, Business Source Complete, EBSCOhost.

150. Gwennap, "Broadcom Gains Ground in the GPS Chip Market."

151. Ludovic Privat, "Samsung Acquires CSR Handset Connectivity and GPS Semiconductor Division," *GPS Business News*, July 18, 2012, http://www.gpsbusinessnews.com/Samsung-Acquires-CSR-Handset-Connectivity-GPS-Semiconductor-Division_a3759.html?preaction=nl&id=13088485&idnl=117462&.

152. Paul Davidson, "Enhanced 911 Calls Still Far from Wide Coverage," *USA Today*, October 25, 2002, LexisNexis Academic.

153. Amy Schatz, "Millions Resist Shift to Mobiles Fit for 911 Calls," *Wall Street Journal*, January 6, 2006, ProQuest.

154. Schatz, "Millions Resist Shift."

155. Erik Amcoff, "Mobile Phones Are Packing More GPS Features," *Wall Street Journal*, March 23, 2005, ProQuest.

156. Denis Dutton, "It's Always the End of the World as We Know It," *New York Times*, January 1, 2010, http://www.nytimes.com/2010/01/01/opinion/01dutton.html?_r=1&pagewanted=all.

157. Steve Lohr, "1/1/00: Technology and 2000—Momentous Relief; Computers Prevail in First Hours of '00," *New York Times*, January 1, 2000.

158. Tyler Hamilton, "Coming Satellite Adjustment a Y2K 'Dress Rehearsal': People Relying on Electronic Tracking Devices at Risk of Going Astray as Orbiting Clocks Are Reset on Saturday," *Globe and Mail* (Canada), August 16, 1999, LexisNexis Academic.

159. William Glanz, "GPS Users May Want to Dust Off Compasses," *Washington Times*, August 19, 1999, LexisNexis Academic.

160. Steve Ranger, "Date Flaw Could Sink Navigation Systems," *Computing*, August 19, 1999, LexisNexis Academic.

161. *IS-GPS-200F*, Global Positioning System Directorate, September 21, 2011, 42, http://www.gps.gov/technical/icwg/#is-gps-200.

162. U.S. Air Force, "Global Positioning System Calendar Faces Rollover," FDCH *Federal Department and Agency Documents*, August 6, 1999, LexisNexis Academic.

163. Glanz, "GPS Users May Want to Dust Off Compasses."

164. Glanz, "GPS Users May Want to Dust Off Compasses."

165. Hamilton, "Coming Satellite Adjustment a Y2K 'Dress Rehearsal.'"

166. "Federal Agencies Warn Drivers, Boaters, Pilots and Hikers of False Readings with Some GPS-based Satellite Navigation Systems," FDCH *Federal Department and Agency Documents*, August 5, 1999, LexisNexis Academic.

167. Hamilton, "Coming Satellite Adjustment a Y2K 'Dress Rehearsal.'"

168. Shahid Ali Kahn, "Over 5,000 to Face Problems with GPS," *Saudi Gazette* via Middle East Newsfile, August 24, 1999, LexisNexis Academic.

169. "End-of-Week Rollover Bug Strikes," *Global Positioning & Navigation News*, August 25, 1999, LexisNexis Academic.

170. Bruce D. Nordwall, "GPS Rollover Passes without Incident," *Aviation Week & Space Technology*, August 30, 1999, LexisNexis.

171. Nordwall, "GPS Rollover Passes without Incident."

172. Paul Daley, "Satellite Guidance Glitch Hits Navy Boat," *The Age* (Melbourne, Australia), August 26, 1999, LexisNexis Academic.

173. Andrew Cornell, "Tokyo Traffic Chaos in GPS Date Rollover," *Australian Financial Review*, August 24, 1999, LexisNexis Academic.

174. Cornell, "Tokyo Traffic Chaos."

175. Doug Struck, "Japan's Cars Lost in Space; Y2K-Like Bug Cuts Some Motorists Off from Navigation System," *Washington Post*, August 24, 1999, LexisNexis Academic.

176. White House, "Statement by the President Regarding the United States' Decision to Stop Degrading Global Positioning System Accuracy," news release, May 1, 2001, National Archives and Records Administration, http://clinton3.nara.gov/WH/EOP/OSTP/html/0053_2.html.

177. Douglas Page, "Rising Star," *Fire Chief*, December 1, 2003, 12, LexisNexis Academic; Sarah Z. Sleeper, "Satellites Help Cut Sept. 11 Cleanup Costs; Positioning Gear Used by Tracking the Location of All Recovery Vehicles, Crews Boosted Efficiency," *Investor's Business Daily*, June 7, 2002, LexisNexis Academic.

178. John Rendleman, "GPS Aids Recovery Effort—Searchers in New York Use Handheld Devices to Keep Track of Evidence," *Information Week*, November 12, 2001, LexisNexis Academic.

179. Page, "Rising Star"; Sleeper, "Satellites Help Cut Sept. 11 Cleanup Costs."

180. Laura Q Hughes, "Breaking Ground; Firms Came to the Rescue at Ground Zero with Handheld Devices, Tracking Systems and Robots, but Involvement Has Come at a Price," *Crain's New York Business*, April 22, 2002, 17, LexisNexis Academic; Sleeper, "Satellites Help Cut Sept. 11 Cleanup Costs."

181. PBS, *Frontline*, "Campaign Against Terror," "On the Ground—War Stories," http://www.pbs.org/wgbh/pages/frontline/shows/campaign/ground/warstories.html (accessed June 28, 2012).

182. Jim Yardley, "FBI Focuses on Navigational Device," *New York Times*, September 29, 2001, http://www.nytimes.com/2001/09/29/national/29FLIG.html?pagewanted=all.

183. "Doesn't Work," *Air Safety Week,* November 5, 2001, LexisNexis Academic.

184. "Doesn't Work."

185. Aaron Brown et al., CNN *NewsNight*, transcript, May 22, 2002, LexisNexis Academic.

186. Greg B. Smith, "Hijacker in City Sept. 10 Used Navigation Tool to Pinpoint WTC Site," *Daily News* (New York), May 22, 2002, LexisNexis Academic; "Atta Thought to Have Cased WTC Before Attack," *Bulletin's Frontrunner*, May 23, 2002, LexisNexis Academic.

187. Indictment, United States v. Zacarias Moussaoui (USDC, E. D. Va., 2001), http://www.justice.gov/ag/moussaouiindictment.htm (accessed June 28, 2012); "Indictment Chronicles Overt Acts That It Says Led to Sept. 11 Attacks," *New York Times*, December 12, 2001, http://www.nytimes.com/2001/12/12/national/12ITEX.html?pagewanted=all.

188. Brief for the United States, United States v. Zacarias Moussaoui, No. 06-4494, Document 209, 41 (4th Cir. August 25, 2008), www.ca4.uscourts.gov/moussaoui4494/pdf/082508brief.pdf.

189. Richard J. Newman and Douglas Pasternak, "A Jekyll and Hyde System," *U.S. News & World Report*, December 24, 2001, 20, LexisNexis Academic.

190. Arik Hesseldahl, "After the Attacks, New Attention on GPS," Forbes .com, October 2, 2001, http://www.forbes.com/2001/10/02/1002gps .html.

191. Hesseldahl, "After the Attacks."

192. Katharine Mieszkowski, "A No-Fly Zone for Terrorism," Salon.com, December 13, 2001, http://www.salon.com/2001/12/13/soft_walls.

193. Mieszkowski, "No-Fly Zone for Terrorism."

194. "GPS: America's 'Maginot Line' in Impending Conflict?" *Global Positioning & Navigation News*, October 3, 2001, LexisNexis Academic.

195. Alan Levin, Marilyn Adams, and Blake Morrison, "Terror Attacks Brought Drastic Action: Clear the Skies," *USA Today*, August 12, 2002, http://www.usatoday.com/news/sept11/2002-08-12-clearskies_x.htm.

196. Levin, Adams, and Morrison, "Terror Attacks Brought Drastic Action."

197. Alan Levin, "For Air Traffic Controllers, a Historic Achievement," *USA Today*, August 12, 2002, http://www.usatoday.com/news/sept11/2002 -08-12-atc_x.htm.

198. Associated Press, "FAA Lifts Restrictions on General Aviation Flights," December 21, 2001, LexisNexis Academic.

199. National Coordination Office for Space-Based Positioning, Navigation and Timing, "Special Notice: September 17, 2001," http://www.gps .gov/systems/gps/modernization/sa/IGEB/.

200. Peter Warner, "U.S. Isolated on Civil GPS Liability Issue," *Global Positioning & Navigation News*, October 17, 2001, LexisNexis Academic.

201. Warner, "U.S. Isolated on Civil GPS Liability Issue."

202. Brian Harvey, *The Rebirth of the Russian Space Program* (Chichester UK: Praxis Publishing, 2007), 129.

203. Bruce D. Nordwall, "It's Never Easy," *Aviation Week & Space Technology*, September 8, 2003, 60, LexisNexis Academic; Charles R. Trimble, letter to the editor, *Foreign Affairs*, September–October 2003, Lexis-Nexis Academic.

204. Kathy Gambrell, "U.S., U.K. Reach Agreement on Satellite Positioning Systems," *Aerospace Daily & Defense Report*, June 22, 2004, LexisNexis Academic.

205. Robert Wall, "Keeping Time: Updated GPS Policy Reflects a Decade of Change in System's Customer Base," *Aviation Week & Space Technology*, December 20, 2004, 34, LexisNexis Academic.

9. Where Are We?

1. Rachel Beck, "Family Fine after Snow Scare," *Corvallis (OR) Gazette-Times*, December 28, 2009, http://www.gazettetimes.com/news/local/7f1f6338-f3d8-11de-8b6c-001cc4c002e0.html; A. K. Dugan, "Lost Family Found on Forest Service Road," *Lebanon (OR) Express*, December 30, 2009, http://lebanon-express.com/news/local/lost-family-found-on-forest-service-road/article_acccef26-f4d2-11de-a852-001cc4c03286.html; Tim Fought, "GPS-Led Travel Goes Amiss; 3 Ore. Parties Rescued," Associated Press, January 2, 2010, LexisNexis Academic.

2. Amy Gilroy, "PNDs Were Hot Black Friday Items," *TWICE*, December 7, 2009, http://www.twice.com/article/438924-PNDs_Were_Hot_Black_Friday_Items.php.

3. Dugan, "Lost Family Found."

4. Beck, "Family Fine after Snow Scare."

5. Fought, "GPS-Led Travel Goes Amiss."

6. Dugan, "Lost Family Found."

7. Fought, "GPS-Led Travel Goes Amiss."

8. "The Air Force Wants You to Stop Blaming GPS Satellites When You Get Lost," Gizmodo.com, December 30, 2009, LexisNexis Academic.

9. "Michael Drives Car into Lake," *The Office*, YouTube.com, http://www.youtube.com/watch?v=BIakZtDmMgo (accessed July 24, 2012).

10. "Technology Blamed as Thousands Get Lost at Newgrange," *Irish Independent,* June 30, 2008, LexisNexis Academic.

11. Al Webb, "Wrong-Way GPS Leading British Astray: Navigation Gizmos Are Unable to Track Narrow Village Lanes, Fields and Streams," *Washington Times*, March 2, 2008, LexisNexis Academic.

12. Margaret Toal, "GPS Gaffe Landlocks Truckers Bound for Port in Residential Neighborhood," *Beaumont (TX) Enterprise*, December 1, 2009, LexisNexis Academic.

13. Associated Press, "Oops! Crew Demolishes Wrong House," June 11, 2009, LexisNexis Academic.

14. Dakota Smith, "GPS Directions to Hollywood Sign Will Change," *Contra Costa (CA) Times*, http://www.contracostatimes.com/california/ci_20186097 (updated March 15, 2012).

15. Charles Malinchak, "GPS Units Sending Rigs down Narrow Roads," *Allentown (PA) Morning Call*, February 4, 2012, LexisNexis Academic.

16. PR Newswire, "Former Cruise Ship Officer M.L. Meier Reveals What Cruise Lines Don't Want You to Know," news release, February 6, 2012, http://media.prnewswire.com/en/jsp/search.jsp?searchtype=full &option=headlines&criteriadisplay=show&resourceid=4976409.

17. Jacquelin Magnay, "London 2012 Olympics: American and Australian Team Buses Get Lost from Heathrow to Olympic Park," *London Telegraph*, July 16, 2012, http://www.telegraph.co.uk/sport/olympics/ 9403023/London-2012-Olympics-American-and-Australian-team -buses-get-lost-from-Heathrow-to-Olympic-Park.html.

18. Avril Ormsby and Paul Sandle, "Olympics—Don't Tweet If You Want TV, London Fans Told," Reuters, July 29, 2012, http://www.reuters .com/article/2012/07/29/oly-twitter-day-idUSL6E8IT2UH20120729.

19. Mark Hinchliffe, "Navigating Through a Lack of Trust," *Hobart Mercury* (Australia), August 14, 2010, LexisNexis Academic.

20. Callie Watson, "Most Lost Motorists Place Blame on Their GPS Units," *Advertiser* (Australia), December 18, 2009, LexisNexis Academic.

21. Michelin, "Survey: Most Drivers Using GPS Say It Has Led Them Astray," news release, June 17, 2013, http://michelinmedia.com/news/ survey-drivers-gps-led-astray/.

22. Allstate.com, "Survey Finds 83 Percent of Men Occasionally Disobey GPS Directions," http://www.allstate.com/insurance-industry-news/ auto-insurance-news/survey-finds-83-percent-of-men-occasionally -disobey-gps-directions-800487019.aspx (accessed July 22, 2012).

23. Barbara Rodriguez, "Fewer Maps Being Printed as Travelers Reach for Technology," *Norfolk Virginian-Pilot*, August 19, 2012, LexisNexis Academic.

24. Preston Sparks, "Demand Remains for Map of State," *Augusta (GA) Chronicle*, January 15, 2009, LexisNexis Academic.

25. James Q. Lynch, "With an Eye on GPS Navigational Unites, Iowa House Cuts Funds for Road Maps," *Cedar Rapids (IA) Gazette*, March 21, 2012, LexisNexis Academic.

26. Rodriguez, "Fewer Maps Being Printed."

27. Richard Willing, "Thieves Find Way to GPS Units in Cars; High-Tech Gear, Parts Increasingly Irresistible," *USA Today*, June 22, 2006, LexisNexis Academic.

28. Kevin Johnson, "Violent Crime Falls in 2008; FBI Also Sees Rise in Burglaries," *USA Today*, September 15, 2009, LexisNexis Academic.

29. Barbara Boyer, "GPS Devices Lure Thieves," *Philadelphia Inquirer*, April 23, 2008, LexisNexis Academic.

30. John Mooney, "Satnavs a Must-Have, Especially Now for Thieves," *London Times*, September 16, 2007, LexisNexis Academic.

31. Cruz Morcillo and Pablo Munoz, "Economic Crisis Reflected in Growing Crime Figures—Spanish Police," *ABC.es* (Madrid), transcript of Spanish newspaper website by BBC Worldwide Monitoring, September 30, 2008, LexisNexis Academic.

32. David Hastie, "GPS Units on the Map for Thieves," *Herald Sun* (Australia), December 7, 2008, LexisNexis Academic.

33. Ernesto Londoño, "Rising Thefts from Cars Have Police on the Alert; Cellphones, Money, GPS Devices Stolen," *Washington Post*, June 21, 2007, LexisNexis Academic.

34. Business Wire, "Factory Installed Automotive Navigation Solutions to Surpass PND Shipments by 2015, Says ABI Research," news release, May 16, 2012, LexisNexis Academic.

35. "Take Me Home . . . Take Everything," *Navigation News*, May–June 2012, 5.

36. Frank Main, "iPhone Theft Suspects Tripped Up by GPS App," *Chicago Sun-Times*, February 18, 2011, http://www.suntimes.com/3890624 -417/iphone-theft-suspects-tripped-up-by-gps-app.html.

37. UPI, "iPhone App Leads Police to Wrong House," February 27, 2012, LexisNexis Academic.

38. Lauren Davidson, "Tracking Apps Not Fail-Safe Protection," *Atlanta Journal Constitution*, July 10, 2012, LexisNexis Academic.

39. Rick W. Sturdevant, "The Socioeconomic Impact of the NAVSTAR Global Positioning System, 1989–2009," paper presented at the American Institute of Aeronautics and Astronautics (AIAA) Space 2009 Conference and Exposition, September 14–17, 2009, Pasadena CA.

40. Sturdevant, "Socioeconomic Impact of the NAVSTAR Global Positioning System."

41. General Dynamics, "AN/PRC-154 Rifleman Radio," http://www.gdc4s .com/content/detail.cfm?item=b8c971d4-9784-41c7-b8e1-1f557f1b2 dod&page=2 (accessed August 3, 2012); Thales Communications Inc., "AN/PRC Rifleman Radio," http://www.thalescomminc.com/ground/ anprc154.asp (accessed August 3, 2012).

42. Business Wire, "Satellites and Computers Track Houston Buses in

Demonstration of Transit Future," news release, October 1, 1990, Lex-isNexis Academic.

43. PR Newswire, "MARCOR Lands Industry's Largest GPS Commercial Contract: Over 1,600 Buses and Cars to Be Equipped with Satellite Tracking System," news release, September 19, 1991, LexisNexis Academic.

44. Sturdevant, "Socioeconomic Impact of the NAVSTAR Global Positioning System."

45. Pham, *Economic Benefits of Commercial GPS Use*, 9.

46. Youtube.com, "AT&T TV Commercial: Genco Services," http://www.youtube.com/watch?v=Z3QHQEXb9Tg&feature=player_embedded (accessed August 20, 2012).

47. Sturdevant, "Socioeconomic Impact of the NAVSTAR Global Positioning System."

48. UPI, "Bank Robbery Suspect Caught on City Bus," May 30, 2012, Lexis-Nexis Academic.

49. Jeff Casale, Mark A. Hoffmann, and Sally Roberts, "Town Can Always Find Jesus Thanks to GPS," *Business Insurance* 42, no. 51 (December 22, 2008), 23, Business Source Complete, EBSCOhost.

50. Tagg website, "Products: First Pet Master Kit—Monthly," http://www.pettracker.com (accessed August 4, 2012).

51. Pham, *Economic Benefits of Commercial GPS Use*, 4; Roger Yu, "Wearable Devices Support Wireless Tracking; Makers Incorporate Sensors, Other Tech," *USA Today*, May 1, 2012, LexisNexis Academic.

52. Alzheimer's Association, "About Comfort Zone," http://www.alz.org/comfortzone/about_comfort_zone.asp (accessed August 11, 2012); Omnilink Systems, "Reliable Tracking Technology for Alzheimer's and Dementia," http://www.omnilink.com/comfort-zone/ (accessed June 16, 2013).

53. AmberWatch Foundation, "AmberWatch GPS," http://www.amberwatchgps.com (accessed August 11, 2012).

54. Aetrex, "Aetrex Navistar GPS Footwear System FAQ," http://www.navistargpsshoe.com/gps-faq (accessed August 11, 2012).

55. Peggy J. Noonan, "Safety Afoot, via GPS, for Alzheimer's Patients," *USA Weekend*, October 14, 2010.

56. Teniska Museet, "100 Innovations: An Exhibition at the National Museum of Science and Technology, Stockholm, Sweden, 2012–2015,"

http://www.100innovationer.com/engelsk/theinnovations/
innovations/gps.341.html (accessed August 11, 2012).

57. Associated Press, "Swedish Preschools Use GPS to Track Children,"
    Greensboro (NC) News & Record, September 25, 2011; Purple Scout.com,
    "Barnkollen (ChildChecker)," http://www.purplescout.com (accessed
    August 4, 2012).

58. Associated Press, "Swedish Preschools Use GPS to Track Children."

59. Revolutionary Tracker, "The RT-01 GPS Watch Locator," http://www
    .revolutionarytracker.com/pro (accessed August 11, 2012).

60. Amber Alert GPS, "Your Child's Personal GPS Locator," http://www
    .amberalertgps.com/product-device (accessed August 11, 2012).

61. USCourts.gov, "GPS: Your Supervising Officer Is Watching You," http://
    www.uscourts.gov/news/TheThirdBranch/07-04-01/GPS_Your
    _Supervising_Officer_is_Watching.aspx (accessed August 4, 2012).

62. Patrick Hyde and Nicole DeJarnatt, "GPS Offender Tracking and the
    Police Officer," *Law Enforcement Technology*, June 2005, LexisNexis
    Academic.

63. "Prosthetic Leg Used to Trick Tracking Tag," UPI *Quirks in the News*,
    August 29, 2011, Newspaper Source Plus, EBSCOhost.

64. Todd Richmond, "Company: Electronic Monitoring Went Down
    across U.S.," Associated Press, October 7, 2010, LexisNexis Academic.

65. Sadie Gurman, "Monitoring Device's Use Questioned," *Pittsburgh
    Post-Gazette*, December 9 2009, LexisNexis Academic; Park Si-soo,
    "Efficacy of Electronic Bracelet Questioned," *Korea Times*, December
    8, 2010, LexisNexis Academic.

66. Justin Fenton, "Mother of Girl Hurt in Shooting Sues; GPS Maker
    Should Have Known Shooter Could Fool It, They Say," *Baltimore Sun*,
    June 14, 2012, LexisNexis Academic.

67. Scott J. Croteau, "GPS Home Confinement under Review; DA Orders
    Study of Monitoring after Defendant's Complaint," *Worcester (MA)
    Telegram & Gazette*, February 16, 2012, LexisNexis Academic.

68. Croteau, "GPS Home Confinement under Review."

69. "GPS-Aided Monitoring of Parolees: No Privacy Issues, Just a Large
    Addressable Market," *Inside GNSS*, May 20, 2012, http://www.inside
    gnss.com/node/3056.

70. "GPS-Aided Monitoring of Parolees."

71. Jim Stark, "GPS Tracking Is the Wave of the Future for Law Enforce-

ment Authorities," *Journal of Counterterrorism & Homeland Security International* 9, no. 1 (Winter 2003), LexisNexis Academic.

72. Peter A. Michel, "Now Is the Right Time to Use Electronic Monitoring," *Corrections*, December 21, 2009, http://www.corrections.com/articles/23004.

73. "GPS-Aided Monitoring of Parolees"; James Kilgore, "Electronic Monitoring: Some Causes for Concern," *Prison Legal News Online*, https://www.prisonlegalnews.org/%28S%28ynfmxhexm1f5vljhnw3zvoub%29%29/displayArticle.aspx?articleid=24254&AspxAutoDetectCookieSupport=1 (accessed August 4, 2012).

74. GEO Group, "The GEO Group Closes $415 Million Acquisition of B.I. Incorporated," news release, February 11, 2011, http://phx.corporate-ir.net/phoenix.zhtml?c=91331&p=irol-newsArticle&ID=1528023&highlight=.

75. GEO Group, "GEO Group Closes $415 Million Acquisition."

76. "Company Sees Prisoner Tracking and Monitoring Niche," *Global Positioning & Navigation News*, May 16, 1996, LexisNexis Academic.

77. "Dmatek Post Strongest Results Yet," *Investors Chronicle*, March 7, 2008, LexisNexis Academic.

78. Nadav Shemer, "Business Briefs," *Jerusalem Post*, March 16, 2011, LexisNexis Academic; 3M, "About 3M Attenti: History," http://www.attentigroup.com/default.asp?PageID=12 (accessed August 11, 2012); 3M, "SEC Filings, 2011 Annual Report," http://media.corporate-ir.net/media_files/irol/80/80574/Annual_Report_2011.pdf (accessed August 11, 2012).

79. Hoovers.com, "Attenti Ltd.: Competitive Landscape," http://subscriber.hoovers.com.proxy067.nclive.org/H/company360/competitiveLandscape.html?companyId=103073000000000 (accessed August 12, 2012).

80. SecureAlert, "SecureAlert Investor Brief," http://securealert.com/Overview (accessed August 12, 2012).

81. David Last, "Silent Witness," *Navigation News*, January–February 2009, 10–13.

82. Ben Hubbard, "Police Turn to Secret Weapon: GPS Device," *Washington Post*, August 13, 2008, LexisNexis Academic.

83. Hubbard, "Police Turn to Secret Weapon."

84. Joan Biskupic, "Court: GPS Tracking Needs a Warrant," *USA Today*,

January 24, 2012, http://www.usatoday.com/USCP/PNI/Nation/
World/2012-01-24-bcUSATSCOTUSGPSUPDT_ST_U.htm.

85. Adam Liptak, "Justices Say GPS Tracker Violated Privacy Rights," *New York Times*, January 23, 2012, http://www.nytimes.com/2012/01/24/us/
police-use-of-gps-is-ruled-unconstitutional.html?_r=1&pagewanted
=all.

86. SupremeCourt.gov, "2011 Term Opinions of the Court, Slip Opinions, Per Curiams (PC), and Original Cases Decrees (D): United States v. Jones," 4, http://www.supremecourt.gov/opinions/11pdf/10-1259.pdf
(accessed August 13, 2012).

87. Aaron Kase, "When Is It Okay to Use GPS Trackers?" blog at Lawyers.
com, February 7, 2012, http://blogs.lawyers.com/2012/02/when-is-it
-okay-to-use-gps-trackers.

88. Kase, "When Is It Okay to Use GPS Trackers?"

89. Alasdair Allan and Pete Warden, "Got an iPhone or 3G iPad? Apple Is Recording Your Moves," *O'Reilly Radar* (blog), April 20, 2011, http://
radar.oreilly.com/2011/04/apple-location-tracking.html.

90. Apple.com, "Apple Q&A on Location Data," news release, April 27, 2011, http://www.apple.com/pr/library/2011/04/27Apple-Q-A-on
-Location-Data.html.

91. John J. Heitmann and Christopher S. Koves, "Senator Leahy Intro-
duces Bill to Update Electronic Communications Privacy Act," blog at AdLawAccess.com, May 23, 2011, http://www.adlawaccess.
com/2011/05/articles/privacy-and-information-securi/senator-leahy
-introduces-bill-to-update-electronic-communications-privacy-act.

92. GovTrack.us, "Congress: Bills: H.R. 1895," http://www.govtrack.us/
congress/bills/112/hr1895 (accessed August 15, 2012).

93. Steve Stecklow and Julia Angwin, "Corporate News: House Releases 'Do Not Track' Bill," *Wall Street Journal*, May 7, 2011, ProQuest.

94. Chaffetz.House.gov, "Chaffetz, Wyden Introduce GPS Act: Bipartisan Legislation Provides Needed Legal Clarity for Use of Geolocation Information," news release, June 15, 2011, http://chaffetz.house.gov/
press-release/chaffetz-wyden-introduce-gps-act-bipartisan-legislation
-provides-needed-legal-clarity.

95. GovTrack.us, "Congress: Bills: H.R. 2168: Geolocational Privacy and Surveillance Act," http://www.govtrack.us/congress/bills/112/hr2168
(accessed August 14, 2012); GovTrack.us, "Congress: Bills: S. 1212:

Geolocational Privacy and Surveillance Act," http://www.govtrack.us/congress/bills/112/s1212 (accessed August 14, 2012).

96. GovTrack.us, "Congress: Bills: S. 1223: Location Privacy Act of 2011," http://www.govtrack.us/congress/bills/112/s1223 (accessed August 14, 2012).

97. John J. Heitmann and Christopher S. Koves, "Flood of Geolocational Privacy Legislation Introduced in June," blog at AdLawAccess.com, June 22, 2011, http://www.adlawaccess.com/2011/06/articles/privacy-and-information-securi/flood-of-geolocational-privacy-legislation-introduced-in-june.

98. Targeted News Service, "Wyden Amendments to Cyber Bill Clarify Rules for GPS Tracking; Seek Privacy Protection in the Cloud," news release, August 2, 2012, LexisNexis Academic.

99. Bryce Bashuk and Greg Piper, "Senate Fails to Advance Cybersecurity Bill," *Communications Daily*, August 3, 2012, LexisNexis Academic.

100. GovTrack.us, "Congress: Bills: H.R. 1312: Geolocational Privacy and Surveillance Act," http://www.govtrack.us/congress/bills/113/hr1312 (accessed June 15, 2013); GovTrack.us, "Congress: Bills: S. 639: Geolocational Privacy and Surveillance Act," http://www.govtrack.us/congress/bills/113/s639 (accessed June 15, 2013).

101. GovTrack.us, "Congress: Bills: H.R. 1312"; GovTrack.us, "Congress: Bills: S. 639."

102. Chaffetz.House.gov, "Chaffetz, Wyden Introduce GPS Act."

103. Kunur Patel, "Know What Your Phone Knows about You? From Location Info to Subscriber Data, Carriers, Others Confront Mobile Privacy," *Advertising Age*, March 7, 2011, LexisNexis Academic.

104. PR Newswire, "Global Location-Based Advertising Market to Reach $6.2 Billion in 2015, Pyramid Research," news release, June 21, 2011, http://media.prnewswire.com/en/jsp/main.jsp?clientid=1098230&option=cip&view=releases&resourceid=4649811.

105. Glen Gibbons, e-mail to Eric Frazier, August 20, 2012.

106. Noam Cohen, "It's Tracking Your Every Move and You May Not Even Know It," *New York Times*, March 26, 2011, http://www.nytimes.com/2011/03/26/business/media/26privacy.html.

107. Eric Lichtblau, "More Demands on Cell Carriers in Surveillance," *New York Times*, July 8, 2012, http://www.nytimes.com/2012/07/09/us/cell-carriers-see-uptick-in-requests-to-aid-surveillance.html?_r=1.

108. Markey.House.gov, "Markey: Law Enforcement Collecting Information on Millions of Americans from Mobile Phone Carriers," news release, July 9, 2012, http://markey.house.gov/press-release/markey-law-enforcement-collecting-information-millions-americans-mobile-phone-carriers.

109. Markey.House.gov, "Markey: Law Enforcement Collecting Information."

110. Lichtblau, "More Demands on Cell Carriers in Surveillance."

111. Lichtblau, "More Demands on Cell Carriers in Surveillance."

112. Jennifer Valentino-DeVries, "Covert FBI Power to Obtain Phone Data Faces Rare Test," *Wall Street Journal*, July 18, 2012, ProQuest.

113. Valentino-DeVries, "Covert FBI Power to Obtain Phone Data."

114. Jennifer Valentino-DeVries and Julia Angwin, "Latest Treasure Is Location Data; as Lawmakers Ready Hearings, Insurers, Car Makers, Even Shopping Malls Seek to Track Customers," *Wall Street Journal*, May 10, 2011, ProQuest.

115. Matthew Lasar, "Dutch Traffic Cops Use TomTom GPS Data to Nail Speeders," *Ars Technica*, April 28, 2011, http://arstechnica.com/tech-policy/2011/04/dutch-traffic-cops-use-tomtom-gps-data-to-nail-speeders.

116. "Apple iPhone Linked to Carrier IQ 'Spyware,'" *London Telegraph*, December 1, 2001, LexisNexis Academic.

117. Jennifer Smith, "Lawyers Get Vigilant on Cybersecurity; Pressure Grows as Mobile Devices, Emails Make Sensitive Data More Vulnerable," *Wall Street Journal*, June 26, 2012, ProQuest.

118. PR Newswire, "AT&T Poll Says Use of Tablets, 4G Devices, GPS Navigation Mobile Apps on the Rise among Small Businesses," news release, February 15, 2012, http://media.prnewswire.com/en/jsp/latest.jsp?resourceid=4997102&access=EH.

119. Business Wire, "Global Market for GPS Navigation and Location–Based Mobile Services to Rise to $15.2. Billion in 2016, a CAGR of 22.7%, According to New Research Report by IE Market Research Corporation," news release, November 21, 2011, LexisNexis Academic.

120. TNS Global, "Two-Thirds of World's Mobile Users Signal They Want to Be Found," news release, April 24, 2012, http://www.tnsglobal.com/press-release/two-thirds-world%E2%80%99s-mobile-users-signal-they-want-be-found.

121. TNS Global, "Two-Thirds of World's Mobile Users."

122. Sarah Perez, "All the Location Apps You Have to Use at the SXSW Royal Rumble," blog at TechCrunch.com, March 6, 2012, http://techcrunch .com/2012/03/06/all-the-location-apps-you-have-to-use-at-the-sxsw -royal-rumble.

123. Gert Ian Spriensma, "Publication: Social Networking Apps," blog at Distimo.com, August 30, 2012, http://www.distimo.com/blog/2012_08 _publication-social-networking-apps.

124. Gert Ian Spriensma, "Publication: Social Networking Apps."

125. Ben Rooney, "Facebook, Groupon Hit New Lows," *The Buzz* (blog), CNN Money, September 4, 2012, http://buzz.money.cnn.com/2012/ 09/04/facebook-groupon-hit-new-lows.

126. AppAppeal.com, "Top 4 Free Location Based Networking Apps," http://www.appappeal.com/apps/location-based-networking (accessed September 5, 2012); Business Wire, "Green Dot to Acquire Loopt," news release, March 9, 2012, http://www.businesswire.com/ news/home/20120309005422/en/Green-Dot-Acquire-Loopt.

127. Elizabeth Woyke, "Patents Suggest 'Indoor Location' the Next Mobile Maps Battleground," *Globe and Mail* (Canada), http://www.theglobe andmail.com/technology/tech-news/patents-suggest-indoor-location -the-next-mobile-maps-battleground/article4248246/?service=print (updated December 29, 2011).

128. Kathleen Hickey, "New Chip Tracks Smart-Phone Locations Down to the Inch, Even Indoors," *Government Computer News*, April 26, 2012, LexisNexis Academic; Woyke, "Patents Suggest 'Indoor Location' the Next Mobile Maps Battleground."

129. Jamie Lendino, "Apple Moves to Kill GPS Devices, Reduce Dependence on Google," *PC Magazine*, June 12, 2012, http://www.pcmag .com/article2/0,2817,2405680,00.asp.

130. Ludovic Privat, "Google Claims 10,000 Indoor Locations Mapped," *GPS Business News*, July 11, 2012, http://www.gpsbusinessnews.com/ Google-Claims-10000-Indoor-Locations-Mapped_a3744.html.

131. Ludovic Privat, "Walgreens: 7,907 Stores Mapped Indoor with Aisle 411," *GPS Business News*, July 23, 2012, http://www.gpsbusinessnews .com/Walgreens-7907-Stores-Mapped-Indoor-with aisle411_a3768 .html?preaction=nl&id=13088485&idnl=117788&; Aisle411.com, "About Aisle411," http://aisle411.com/about-aisle411 (accessed September 6, 2012).

132. PR Newswire, "Valpak Launches Location Based Coupons with junaio Augmented Reality Partner," news release, March 29, 2011, http://www.prnewswire.com/news-releases/valpak-launches-location-based-coupons-with-junaio-augmented-reality-partner-118831664.html.

133. Zachary Lutz, "IBM Labs Pitches the Future of Augmented Reality Shopping with Mobile App Prototype," Engadget.com, July 2, 2012, http://www.engadget.com/2012/07/02/ibm-augmented-reality-shopping-app.

134. MarketsandMarkets, "Indoor Location Market Worth $2.6 Billion by 2018," news release, http://www.marketsandmarkets.com/PressReleases/indoor-location.asp (accessed June 17, 2013).

135. Pham, *Economic Benefits of Commercial GPS Use*, 3.

136. Pham, *Economic Benefits of Commercial GPS Use*, 3.

137. PR Newswire, "MarketsandMarkets: Global GPS Market Worth $26.67 Billion by 2016."

138. Mobile 311, "About Us," http://www.mobile311.com/AboutUs.aspx (accessed August 18, 2012).

139. CityofBoston.gov, "Citizens Connect: Making Boston More Beautiful," http://www.cityofboston.gov/doit/apps/citizensconnect.asp (accessed August 18, 2012).

140. CityofBoston.gov, "Street Bump: Help Improve Your Streets," http://www.cityofboston.gov/DoIT/apps/streetbump.asp (accessed August 18, 2012).

141. U.S. Census Bureau, "Data Protection and Privacy Policy," http://www.census.gov/privacy/data_protection/gps_coordinates.html (accessed August 24, 2012).

142. Jeffrey Collins, "N.C, S.C. State Line Moving South; 'It's Going to Shut This Business,'" *Charlotte (NC) Observer*, March 23, 2012, http://www.charlotteobserver.com/2012/03/22/3118336/nc-sc-state-line-isnt-where-folks.html; Julie Rose, "Boundary Commission Tackles Boundary Puzzles," WFAE 90.7 FM, March 26, 2012, http://www.wfae.org/wfae/19_100_0.cfm?action=display&id=8458.

143. Barbara Carton, "Farmers Begin Harvesting Satellite Data to Boost Yields—GPS Technology Can Micromanage Fields Right Down to the Thistle Patch," *Wall Street Journal*, July 11, 1996, LexisNexis Academic.

144. Carton, "Farmers Begin Harvesting Satellite Data."

145. Pham, *Economic Benefits of Commercial GPS Use*, 6–7.

146. Pham, *Economic Benefits of Commercial GPS Use*, 6–7.

147. Pham, *Economic Benefits of Commercial GPS Use*, 6–7.

148. American Soybean Association, "Ag Groups Say U.S. Agriculture Would Be Gravely Harmed by LightSquared's Plans," news release, March 14, 2012, http://www.soygrowers.com/newsroom/releases/ 2012_releases/r031412b.htm.

149. Andrew Mole, "Precision Is Farming Future," *Weekly Times* (Australia), June 20, 2012, LexisNexis Academic.

150. Sam Barnes, "GPS to Achieve Centimeter Accuracy," *Louisiana Contractor*, November 1999, LexisNexis Academic.

151. Curt Bennink, "3-D Transforms Milling & Paving," *Asphalt Contractor*, November 2010, LexisNexis Academic.

152. Jerry McGraner and Tom Flannagan, "Lengthening a Landfill's Life," *Waste Age*, March 1, 2011, LexisNexis Academic.

153. Dan Brown, "Seeing Beneath the Surface: A 3-D GPS System Takes the Guesswork Out of Underwater Excavation on Ohio River Project," *SitePrep*, January 2012, LexisNexis Academic.

154. Pham, *Economic Benefits of Commercial GPS Use*, 8.

155. Gethin Wyn Roberts, Xiaolin Meng, and Chris Brown, "When Bridges Move: GPS-based Deflection Monitoring," *Sensors*, April 2006, LexisNexis Academic.

156. "Builders Start Assembly of Bridge Satellite Monitoring System," *ITAR-TASS*, November 1, 2011, LexisNexis Academic; Miriam Elder, "Russian City of Vladivostok Unveils Record-Breaking Suspension Bridge," *Guardian*, July 2, 2012, http://www.guardian.co.uk/world/2012/jul/02/ russian-vladivostok-record-suspension-bridge.

157. Xinhua, "Satellite System for Seismic Survey Work," BBC Summary of World Broadcasts, June 15, 1988, LexisNexis Academic.

158. Gary Robbins, "Quakes: Space Used to Pinpoint Ground Movement on Earth," *Orange County (CA) Register*, April 23, 1991, LexisNexis Academic.

159. Yomiuri Shimbun, "More Sites Considered to Monitor Earthquakes," *Daily Shimbun* (Japan), January 21, 1996, LexisNexis Academic; Eric Talmadge, "One Year after Kobe, Researchers Labor at Quake Prediction," Associated Press, January 16, 1996, LexisNexis Academic.

160. Betsy Mason, "Chile Earthquake Moved Entire City 10 Feet to the West," blog at Wired.com, March 8, 2010, http://www.wired.com/

wiredscience/2010/03/chile-earthquake-moved-entire-city-10-feet
-to-the-west.

161. Mason, "Chile Earthquake Moved Entire City."

162. James Dacey, "Rapidly Spotting Major Earthquakes Using GPS," *Phys-ics World*, April 9, 2011, http://physicsworld.com/cws/article/
news/2012/apr/09/rapidly-spotting-major-earthquakes-using-gps;
"GPS Could Warn of Tsunamis," *Investor's Business Daily*, April 26, 2012,
Business Source Complete, EBSCOhost.

163. PR Newswire, "NASA Tests GPS Monitoring System for Big U.S. Earth-quakes," news release, April 24, 2012, http://www.prnewswire.com/
news-releases/nasa-tests-gps-monitoring-system-for-big-us-earth
quakes-148720795.html.

164. Henry Kenyon, "NASA Looks to Space for Better Earthquake Data,"
Government Computer News, April 27, 2012, LexisNexis Academic.

165. Dacey, "Rapidly Spotting Major Earthquakes Using GPS."

166. Edward Powers, "Applications of GPS Provided Time and Frequency
and Future," paper presented at the Tenth Meeting of the National PNT
Advisory Board, Sheraton Pentagon City, Arlington VA, August 14,
2012, http://www.gps.gov/governance/advisory/meetings/2012-08/
powers.pdf.

167. Spectracom Corp., "GPS Clock Synchronization," http://www.spectra
comcorp.com/Solutions/Applications/GPSClockSynchronization/
tabid/100/Default.aspx (accessed April 4, 2013).

168. Symmetricom Inc., "Symmetricom Delivers Precise Time to the Next
Generation Smart Grid," news release, January 28, 2013, http://www
.symmetricom.com/company/news-and-events/press-room/index
.cfm?releaseID=1778484.

169. Powers, "Applications of GPS Provided Time and Frequency and Future."

170. Pham, *Economic Benefits of Commercial GPS Use*, 9–10.

171. U.S. General Accountability Office, *Global Positioning System: Signifi-cant Challenges in Sustaining and Upgrading Widely Used Capabilities*,
GAO-09-670T (Washington DC, May 7, 2009), 3–4, http://www.gao
.gov/assets/130/122502.pdf.

172. U.S. General Accountability Office, *Global Positioning System*.

173. Catherine Rampell, "Nips and Tucks to the Federal Budget," *Economix*
(blog), *New York Times*, May 7, 2009, http://economix.blogs.nytimes
.com/2009/05/07/nips-and-tucks-to-the-federal-budget.

174. William Jackson, "DOD Faces Tough Hurdles in Maintaining, Upgrading an Aging GPS," *Government Computer News*, May 8, 2009, Lexis-Nexis Academic.

175. Gizmodo.com, "GPS Accuracy Might Be Less Accurate in 2010," blog, May 14, 2009, LexisNexis Academic; Engadget.com, "GPS System Might Begin to Fail in 2010, Government Accounting Office Warns," blog, May 19, 2009, LexisNexis Academic.

176. "GPS System 'Close to Breakdown,'" *Guardian Unlimited*, May 19, 2009, LexisNexis Academic; "Satnav Crisis: So Where the Hell Are We?" *mX* (Melbourne, Australia), May 20, 2009, LexisNexis Academic.

177. Ivan Petrov, "Russia's GLONASS to Profit from GPS Problems," *RusData DiaLine*–Russian Press Digest, May 22, 2009, LexisNexis Academic.

178. States News Service, "Space Command Official Tweets on GPS," news release, May 21, 2009, LexisNexis Academic.

179. States News Service, "Space Command Official Tweets on GPS."

180. YouTube.com, "2SOPS Keeps GPS Flying," http://www.youtube.com/watch?v=unFVspzSIr0&lr=1 (accessed August 17, 2012).

181. Callan James, "Fixing GPS," *Avionics Today*, October 1, 2009, http://www.aviationtoday.com/av/issue/cover/Fixing-GPS_35197.html.

182. David Goldstein, "GPS IIR-20 (SVN-49) Information," presentation to 2010 International Technical Meeting (ITM) of the Institute of Navigation (ION), January 26, 2010, Anaheim CA, http://www.gps.gov/multimedia/presentations/2010/01/ITM/goldstein.ppt; Alan Cameron, "Here's What's Up with SVN49," *GPS World*, November 23, 2009, http://www.gpsworld.com/gnss-system/heres-whats-up-with-svn49-9166.

183. Goldstein, "GPS IIR-20 (SVN-49) Information"; Cameron, "Here's What's Up with SVN49."

184. GPS.gov, "Space Segment: Current and Future Satellite Generations: GPS Block IIR(M)," http://www.gps.gov/systems/gps/space (accessed August 24, 2012); U.S. Coast Guard Navigation Center, "GPS Constellation Status for 08/24/2012," http://www.navcen.uscg.gov/?Do=constellationStatus (accessed August 24, 2012).

185. Iridium Communications, "Iridium/Boeing Team Completes High Integrity GPS Milestones," news release, July 13, 2009, http://investor.iridium.com/releasedetail.cfm?ReleaseID=429130; DefenseIndustry Daily.com, "High Integrity GPS/iGPS: Boeing's Iridium Ace Card," July

13, 2013, http://www.defenseindustrydaily.com/Boeing-Wins-RD
-Contract-for-High-Integrity-GPS-05000.

186. Jeffrey Hill, "Boeing Brings Iridium Capability to U.S. GPS System," *Satellite Today*, October 4, 2012, http://www.satellitetoday.com/military/headlines/Boeing-Brings-Iridium-Capability-to-U-S-GPS
-System_39624.html.

187. Amy Butler and Michael Mecham, "GPS IIF Fix Needed," *Aviation Week & Space Technology*, September 6, 2010, 35, LexisNexis Academic; Titus Ledbetter III, "Software Issues Found with GPS IIF-1, Nuclear Detection Payload Harmed," *Inside the Air Force*, September 10, 2010, LexisNexis Academic.

188. Titus Ledbetter III, "GPS II-F Software Fixes to be Deployed On-orbit Early Next Year," *Inside the Air Force*, September 24, 2010, LexisNexis Academic.

189. Dan Elliott, "Glitch Shows How Much U.S. Military Relies on GPS," Associated Press, June 1, 2010, LexisNexis Academic.

190. Titus Ledbetter III, "GPS IIF-2 Launch Delayed to Prolong Anomaly Investigation of Program," *Inside the Air Force*, June 10, 2011, LexisNexis Academic; Amy Butler, "USAF Eyes Fix for GPS IIF Electrical Problem," *Aerospace Daily & Defense Report*, June 21, 2011, LexisNexis Academic.

191. "More GPS Brownouts Ahead," *Farm Industry News*, February 25, 2010, http://farmindustrynews.com/precision-farming/more-gps-brownouts
-ahead; James, "Fixing GPS."

192. Harvey, *Rebirth of the Russian Space Program*, 127–29.

193. Jonathan Beaty and S. C. Gwynne, *The Outlaw Bank: A Wild Ride into the Heart of BCCI* (New York: Random House, 1993), 251–52.

194. Beaty and Gwynne, *Outlaw Bank*.

195. ITAR-TASS, "Yeltsin Clears GLONASS Space Navigation System for Civilian Use," BBC Summary of World Broadcasts, February 22, 1999, LexisNexis Academic; "Cabinet Approves Fund for GLONASS," *Moscow Times*, August 3, 2001, LexisNexis Academic.

196. Harvey, *Rebirth of the Russian Space Program*, 129–30.

197. Harvey, *Rebirth of the Russian Space Program*, 129–30.

198. "GLONASS Plans 30 Satellites, Complete Augmentation System and Improved OCX by 2020," *Inside GNSS*, March–April 2012, http://www
.insidegnss.com/node/3010.

199. "GLONASS Plans 30 Satellites."

200. Harvey, *Rebirth of the Russian Space Program*, 130.

201. DefenseWorld.net, "Russia Offers India Participation in GLONASS GPS Satellite Project," July 18, 2012, http://www.defenseworld.net/go/ defensenews.jsp?id=7218&h=Russia%20Offers%20India%20 Participation%20In%20GLONASS%20GPS%20Satellite%20Project.

202. "Russia Threatens Device Makers to Force GLONASS Acceptance," *Navigation News*, July–August 2010, 4.

203. Arthur Thayer, "West Europe, US Propose Rival Satellite Navigation Systems," *Christian Science Monitor*, December 2, 1983, LexisNexis Academic.

204. Thayer, "West Europe, US Propose Rival Satellite Navigation Systems."

205. Brian Evans, "Setting a Course: Beyond GPS," *Avionics*, May 1, 2006, http://www.aviationtoday.com/av/issue/feature/Setting-a-Course -Beyond-GPS_896.html.

206. Iannotta, "Europe Weighs Building Its Own GPS," 34; Gambrell, "U.S., U.K. Reach Agreement on Satellite Positioning Systems."

207. Dan Levin, "Chinese Square Off with Europe in Space," *New York Times*, March 23, 2009, http://www.nytimes.com/2009/03/23/ technology/23iht-galileo23.html?pagewanted=all.

208. Stephen Chen, "Satellite Launch 'Overrides' EU Deal," *South China Morning Post*, April 16, 2009, LexisNexis Academic.

209. Chen, "Satellite Launch 'Overrides' EU Deal"; Levin, "Chinese Square Off with Europe in Space."

210. Levin, "Chinese Square Off with Europe in Space."

211. Minnie Chan, "'Unforgettable Humiliation' Led to Development of GPS Equivalent," *South China Morning Post*, November 13, 2009, Lexis-Nexis Academic.

212. Xin Dingding and Cheng Yingqi, "Beidou Helps Put Region on the Map," *China Daily*, December 28, 2012, http://www.chinadaily.com .cn/china/2012-12/28/content_16063081.htm.

213. Xinhuanet.com, "Beidou Navigation System Test Network Set for 2015," August 3, 2012, http://news.xinhuanet.com/english/sci/201208/ 03/c_131759872.htm.

214. Callan James, "Satellite Constellations," *Avionics Today*, March 1, 2012, http://www.aviationtoday.com/av/issue/feature/Satellite -Constellations_75794.html; Eric Gakstatter, "Everything Else but GPS: How GLONASS, Galileo, and Compass Will Affect High-Precision

Users," *GPS World* webinar, March 15, 2012; Xinhuanet.com, "Beidou Navigation System Test Network Set for 2015"; Xinhuanet.com, "Tech Base Established for Beidou Navigation System," August 23, 2012, http://news.xinhuanet.com/english/sci/201208/23/c_131801646.htm.

215. Wang Xiaodong, "Beidou Navigates Path to Larger Market Share," *China Daily*, http://usa.chinadaily.com.cn/business/201206/04/content_15468462.htm (updated June 4, 2012); Xinhuanet.com, "Beidou Navigation System Test Network Set for 2015."

216. Xiaodong, "Beidou Navigates Path to Larger Market Share."

217. European Space Agency, "Fact Sheet: What Is Galileo," http://download.esa.int/docs/Galileo_iov_Launch/Galileo_factsheet_20120321.pdf (accessed August 28, 2012).

218. European Space Agency, "Fact Sheet: What is Galileo."

219. European Space Agency, "Galileo Technology Developments," http://www.esa.int/esaNA/SEMP0I5V9ED_galileo_0.html (accessed August 28, 2012).

220. European GNSS Agency, "Galileo Services," http://www.gsa.europa.eu/galileo/services (accessed August 28, 2012).

221. European Space Agency, "Galileo Specifications: Galileo Navigation Signals and Frequencies," http://www.esa.int/esaNA/SEM86CSMD6E_galileo_0.html (updated August 16, 2007).

222. Dee Ann Divis, "British Military Claims Patent on GPS, Galileo Civil Signal Structure," *Inside GNSS*, April 27, 2012, http://insidegnss.com/node/3040.

223. Divis, "British Military Claims Patent on GPS."

224. Dee Ann Divis, "USPTO Nears Approval of Troubling British Patent on New GPS Civil Signal," *Inside GNSS*, May–June 2012, http://www.insidegnss.com/node/3074.

225. "UK to Drop Key GNSS Signal Patent Claim," *Inside GNSS*, November–December 2012, http://online.qmags.com/GNSS1112?fs=2&pg=13&mode=2#pg19&mode2.

226. "Galileo Needs More Cash as UK Chancellor Wields Axe," *Navigation News*, July–August 2010, 4.

227. European Space Agency, "Fact Sheet: What Is Galileo."

228. Dugie Standeford, "EU Lawmakers Begin Talks on New Galileo Financing, Governance as Old Money Worries Remain," *Satellite Week*, March 26, 2012, LexisNexis Academic.

229. Alison Abbott, "Europe's Leaders Slash Proposed Research Budget," Nature.com, February 8, 2013, http://www.nature.com/news/europe -s-leaders-slash-proposed-research-budget-1.12403.

230. European Commission, "Enterprise and Industry: Satellite Navigation: Galileo: Benefits," http://ec.europa.eu/enterprise/policies/satnav/ galileo/index_en.htm (updated February 2, 2012).

231. European Commission, "Enterprise and Industry: Satellite Navigation: Galileo: Benefits."

232. PR Newswire, "European Children Will Name Galileo Satellites Constellation," news release, September 1, 2011, http://media.prnewswire .com/en/jsp/latest.jsp?resourceid=4719907&access=EH.

233. Navipedia.net, "QZSS," http://www.navipedia.net/index.php/QZSS (accessed August 30, 2012).

234. "Japan Aims at 4-Satellite QZSS by Decade's End, India Plans a GAGAN Launch Later This Year," *Inside GNSS*, March–April 2012, http://www .insidegnss.com/node/3014; Navipedia.net, "QZSS."

235. "Japan Aims at 4-Satellite QZSS."

236. Navipedia.net, "QZSS."

237. James, "Satellite Constellations."

238. Navipedia.net, "MSAS General Introduction," http://www.navipedia .net/index.php/MSAS_General_Introduction (accessed August 30, 2012).

239. Navipedia.net, "MSAS General Introduction."

240. Express News Service, "Navigational Satellite Launch Next Year," *New Indian Express*, July 23, 2012, http://newindianexpress.com/cities/ bangalore/article574304.ece.

241. Huma Siddiqui, "India, Russia to Sign Pact to Launch Satellites," *Indian Express*, December 20, 2010, http://www.indianexpress.com/ news/india-russia-to-sign-pact-to-launch-satellites/726745/2.

242. Mandar Chitre and Adam Halliday, "Scientists Excited about India's Own GPS," *Indian Express*, May 20, 2012, http://www.indianexpress .com/news/scientists-excited-about-indias-own-gps/951413/.

243. Navipedia.net, "IRNSS," http://www.navipedia.net/index.php/ IRNSS#cite_note-IRNSS_WIKI-0 (accessed August 20, 2012); Indian Space Research Organisation, "Future Programme: IRNSS-1," http:// www.isro.org/scripts/futureprogramme.aspx (accessed August 30, 2012).

244. Parimal Majithiya, Kriti Khatri, and J. K. Hota, "Indian Regional Navigation Satellite System: Correction Parameters for Timing Group Delays," *Inside GNSS*, January–February 2001, http://www.insidegnss.com/node/2429.

245. Majithiya, Khatri, and Hota, "Indian Regional Navigation Satellite System."

246. Navipedia.net, "GAGAN," http://www.navipedia.net/index.php/GAGAN (accessed August 30, 2012); "GAGAN Project of AAI—Final System Acceptance Test Completed in Bengaluru," *CARGOtalk*, August 8, 2012, http://issuu.com/ashishcb/docs/cargo_aug12/39#.

247. "Japan Aims at 4-Satellite QZSS"; India Space Research Organization website, "Future Programme: Satellite Navigation: GAGAN," http://www.isro.org/scripts/futureprogramme.aspx (accessed April 4, 2013).

248. "Iran Unveils Domestically Manufactured Satellite Navigation System," *Tehran Times*, May 30, 2012, http://www.tehrantimes.com/politics/98404-iran-unveils-domestically-manufactured-satellite-navigation-system.

249. "Iran to Send Satellite into Space Concurrent with Nuclear Talks Day," *Mehr News*, May 14, 2012, http://www.mehrnews.com/en/NewsDetail.aspx?pr=s&query=Fajr%20&NewsID=1602440.

250. Global Security Newswire, "Iran Postpones Satellite Deployment," May 29, 2012, http://www.nti.org/gsn/article/iran-postpones-spacecraft-deployment.

251. Adam M. Brandenburger and Barry J. Nalebuff, *Co-opetition* (New York: Doubleday, 1996), 4, 267. Ray Noorda, founder of the networking company Novell, coined the word *co-opetition* in a December 1993 article in *Electronic Business Buyer*.

252. PR Newswire, "U.S. Air Force Awards Lockheed Martin Contract for Third and Fourth GPS III Satellites," news release, January 12, 2012, http://media.prnewswire.com/en/jsp/search.jsp?searchtype=full&option=headlines&criteriadisplay=show&resourceid=4929218.

253. GPS.gov, "Space Segment: GPS Block IIF," http://www.gps.gov/systems/gps/space/#IIF (accessed August 24, 2012); Gabe Starosta, "Boeing Projecting Completion of GPS IIF Manufacturing by May 2013," *Inside the Air Force*, July 6, 2012, LexisNexis Academic.

254. GPS.gov, "Space Segment: GPS Block IIF"; Starosta, "Boeing Projecting Completion of GPS IIF Manufacturing."

255. Gakstatter, "Everything Else but GPS."

256. Willard Marquis and Michael Shaw, "GPS III: Bringing New Capabilities to the Global Community," *Inside GNSS*, September–October 2011, http://www.insidegnss.com/node/2746; Lockheed Martin, "Lockheed Martin Competes Functional Testing of First GPS II Satellite Bus Electronics Systems," news release, June 5, 2013, http://www.lockheedmartin.com/us/news/press-releases/2013/june/0605-ss-gps.html.

257. Titus Ledbetter III, "Air Force to Move New Space Launch Entrant Rules toward NASA Vision," *Inside the Air Force*, May 20, 2011, LexisNexis Academic; Loren B. Thompson, "Battle Brewing over Future of GPS Constellation," commentary release, States News Service, September 2, 2011, LexisNexis Academic.

258. U.S. General Accountability Office, *Defense Acquisitions: Assessments of Selected Weapons Programs*, GAO-13-294SP (Washington DC, March 2013), 74, http://www.gao.gov/assets/660/653379.pdf.

259. U.S. General Accountability Office, *Defense Acquisitions*, 72.

260. Business Wire, "SkyTerra Communications, Inc. Stockholders Approve Merger with Harbinger," news release, March 22, 2010, LexisNexis Academic.

261. FCC.gov, "Mobile Satellite Ventures Subsidiary LLC: Granted Authority to MSV to Provide ATC Service," 36, http://hraunfoss.fcc.gov/edocs_public/attachmatch/DA-04-3553A1.pdf (accessed September 2, 2012).

262. Gustav Sandstrom, "Nokia Siemens to Build $7 Billion U.S. Network," *Wall Street Journal*, July 20, 2010, ProQuest; GPS.gov, "Reference Documents Related to LightSquared and GPS: November 18, 2010, LightSquared Request for Modification of ATC Authority, or Waiver of Gating Criteria," http://licensing.fcc.gov/myibfs/download.do?attachment_key=852869 (accessed September 2, 2012).

263. Dee Ann Divis, "GPS Community Confronts LightSquared Move into L1 Spectrum," *Inside GNSS*, March–April 2011, http://www.insidegnss.com/node/2508.

264. GPS.gov, "National PNT Advisory Board Recommendations: Letter to FCC Regarding LightSquared," August 3, 2011, http://www.gps.gov/governance/advisory/recommendations/2011-08-lightsquaredlettertofcc.pdf; FCC.gov, "LightSquared Subsidiary LLC: DA 11-133," http://hraunfoss.fcc.gov/edocs_public/attachmatch/DA-11-133A1.pdf (accessed September 2, 2012).

265. Divis, "GPS Community Confronts LightSquared Move."

266. Divis, "GPS Community Confronts LightSquared Move."

267. John Aloysius Farrell and Fred Schulte, "Politically Connected Light-Squared Pushes Wireless Internet Plan despite GPS Concerns," Center for Public Integrity, http://www.publicintegrity.org/2011/07/19/5253/politically-connected-lightsquared-pushes-wireless-internet-plan-despite-gps (updated October 6, 2011).

268. Tim Warren and Adam Bender, "Senators Voice Concerns over Potential LightSquared Interference," *Communications Daily*, April 18, 2011, LexisNexis Academic; Turner.House.gov, "House Passes Turner Language in NDAA which Protects Military's Use of GPS," http://turner.house.gov/news/documentsingle.aspx?DocumentID=272611 (accessed September 3, 2012).

269. PR Newswire, "'Coalition to Save Our GPS' Launched," news release, March 10, 2010, http://www.prnewswire.com/news-releases/coalition-to-save-our-gps-launched-117737113.html.

270. Farrell and Schulte, "Politically Connected LightSquared Pushes Wireless Internet Plan."

271. Josh Smith, "Obama Budget Targets LightSquared," *National Journal*, February 13, 2012, http://techdailydose.nationaljournal.com/2012/02/obama-budget-would-target-ligh.php.

272. Greg Bensinger, "LightSquared Defends Technology in National Ads," *Wall Street Journal*, September 26, 2011, http://online.wsj.com/article/SB10001424052970204831304576594822726900418.html; PR Newswire, "LightSquared Forms Rural Initiative to Ensure LightSquared and GPS Co-existence," news release, July 7, 2011, http://www.prnewswire.com/news-releases/lightsquared-forms-rural-initiative-to-ensure-lightsquared-and-gps-co-existence-125137189.html.

273. FCC.gov, "Final Report 6/30/2011 LightSquared Subsidiary LLC," http://apps.fcc.gov/ecfs/document/view?id=7021690471 (accessed September 4, 2012); PR Newswire, "LightSquared Solution to GPS Issue Will Clear Way for Nationwide 4G Network," news release, June 20, 2011, http://www.prnewswire.com/news-releases/lightsquared-solution-to-gps-issue-will-clear-way-for-nationwide-4g-network-124198384.html.

274. Reuters.com, "From LightSquared: GPS Industry's Failure to Comply with Department of Defense and International Standards for GPS

Receivers Cause of Interference," news release, August 11, 2011, http://www.reuters.com/article/2011/08/11/idUS235128+11-Aug-2011+PRN20110811.

275. FCC.gov, "Status of Testing in Connection with LightSquared's Request for ATC Commercial Operating Authority," public notice, September 13, 2011, http://hraunfoss.fcc.gov/edocs_public/attachmatch/DA-11-1537A1.pdf.

276. PR Newswire, "Statement from Curtis Lu, General Counsel of Light-Squared," news release, October 1, 2011, http://www.prnewswire.com/news-releases/statement-from-curtis-lu-general-counsel-of-light squared-130921943.html; PR Newswire, "Statement by Jeff Carlisle, Executive Vice President for Regulatory Affairs and Public Policy at LightSquared," news release, October 27, 2011, http://www.prnewswire.com/news-releases/statement-by-jeff-carlisle-executive-vice-president-for-regulatory-affairs-and-public-policy-at-lightsquared-132733343.html.

277. PR Newswire, "Former FCC Chief Engineer and LightSquared Question Validity of Test Results Rigged by GPS Industry Insiders," news release, January 18, 2012, http://www.prnewswire.com/news-releases/former-fcc-chief-engineer-and-lightsquared-question-validity-of-test-results-rigged-by-gps-industry-insiders-137565598.html.

278. NTIA.gov, "NTIA Letter to the FCC Regarding LightSquared Request," September 13, 2011, http://www.ntia.doc.gov/files/ntia/publications/ntia_lettertofcc_09132011.pdf.

279. FCC.gov, "Statement from FCC spokesperson Tammy Sun on Letter from NTIA Addressing Harmful Interference Testing Conclusions Pertaining to LightSquared and Global Positioning Systems," news release, February 14, 2012, http://hraunfoss.fcc.gov/edocs_public/attachmatch/DOC-312479A1.pdf.

280. PR Newswire, "LightSquared Response to FCC Public Notice," news release, February 15, 2012, http://www.prnewswire.com/news-releases/lightsquared-response-to-fcc-public-notice-139393213.html.

281. Phil Goldstein, "Harbinger, LightSquared Face Investor Lawsuit, $56 Million Payment," FierceWireless.com, February 21, 2012, http://www.fiercewireless.com/story/harbinger-lightsquared-face-investor-lawsuit-56m-payment/2012-02-21.

282. Mike Dano, "LightSquared's Ahuja Jumps Ship, Company Remains Committed to Network Buildout," FierceWireless.com, February 28,

2012, http://www.fiercewireless.com/story/lightsquareds-ahuja
-jumps-ship-company-remains-committed-network-buildout/
2012-02-28.

283. Sprint Nextel, "Sprint Elects to Terminate Spectrum Hosting Agreement with LightSquared," news release, March 16, 2012, http://news room.sprint.com/article_display.cfm?article_id=2211; Business Wire, "LightSquared Implements Voluntary Chapter 11 Restructuring," news release, May 14, 2012, http://www.businesswire.com/news/home/20120514006605/en/LightSquared-Implements-Voluntary-Chapter-11-Restructuring.

284. GPS.gov, "National PNT Advisory Board Recommendations: White Paper on GPS Jamming," November 4, 2010, http://www.gps.gov/governance/advisory/recommendations/2010-11-jammingwhitepaper.pdf.

285. Pham, *Economic Benefits of Commercial GPS Use*, 2.

286. Pham, *Economic Benefits of Commercial GPS Use*, 15.

287. Dee Ann Divis, "PNT Advisory Board Seeks Details on Economic Benefits of GPS," *Inside GNSS*, August 20, 2012, http://insidegnss.com/node/3170.

10. Going Forward

1. Henry Petroski, *Pushing the Limits: New Adventures in Engineering* (New York: Vintage, 2005), 255.

2. Marquis and Shaw, "GPS III."

3. Marquis and Shaw, "GPS III."

4. Marquis and Shaw, "GPS III."

5. Marquis and Shaw, "GPS III."

6. U.S. General Accountability Office, DOD *Faces Challenges in Fully Realizing Benefits of Satellite Acquisition Improvements*, GAO-12-563T (Washington DC, March 21, 2012), 24, http://www.gao.gov/products/GAO-12-563T.

7. Lockheed Martin, "Global Positioning System: Lockheed Martin's GPS Heritage," http://www.lockheedmartin.com/us/products/gps.html (accessed September 10, 2012).

8. Lockheed Martin, "U.S. Air Force Awards Lockheed Martin Contract for Third and Fourth GPS III Satellites," news release, January 12, 2012, http://www.lockheedmartin.com/us/news/press-releases/2012/january/0112_ss_gps.html.

9. U.S. Air Force, "GPS III Fact Sheet," http://www.losangeles.af.mil/library/factsheets/factsheet_print.asp?fsID=18830&page=1 (accessed September 10, 2012); NASA.gov, "Search and Rescue (SAR)/Distress Alerting Satellite System (DASS)," https://www.spacecomm.nasa.gov/spacecomm/programs/search_and_rescue.cfm (accessed September 10, 2012).

10. Marquis and Shaw, "GPS III."

11. Marquis and Shaw, "GPS III."

12. Dee Ann Divis, "Air Force Proposes Dramatic Redesign for GPS Constellation," *Inside GNSS*, http://www.insidegnss.com/node/3569 (updated June 3, 2013).

13. *IS-GPS-200F*, Global Positioning System Directorate, 4.

14. David Last and Sally Basker, "Bread of Heaven," *Navigation News*, March–April 2012, 11-13.

15. GPS.gov, "National PNT Advisory Board Recommendations: White Paper on GPS Jamming."

16. Callan James, "Precision Approaches," *Avionics*, November 2011, 26-31.

17. GPS.gov, "National PNT Advisory Board Recommendations: White Paper on GPS Jamming."

18. James, "Precision Approaches"; GPS.gov, "National PNT Advisory Board Recommendations: White Paper on GPS Jamming."

19. Federal Communications Commission, *FCC Encyclopedia*, s.v. "Jammer Enforcement," http://www.fcc.gov/encyclopedia/jammer-enforcement (accessed September 15, 2012).

20. Steve Herman, "North Korea Appears Capable of Jamming GPS Receivers," *Voice of America News*, October 7, 2010, LexisNexis Academic.

21. Kim Young-jin, "NK Jams GPS Signals to Disrupt Exercises," *Korea Times*, March 6, 2011, LexisNexis Academic.

22. Yonhap News Agency (Seoul), "North Korea Continues to Disrupt South's Air, Maritime Traffic," BBC Worldwide Monitoring, May 14, 2012, LexisNexis Academic.

23. Song Sang-ho, "S. Korea, China to Cooperate on N.K. GPS Jamming," *Korea Herald*, May 14, 2012, LexisNexis Academic.

24. UPI.com, "North Korea Denies Jamming GPS Signals," May 18, 2012, http://www.upi.com/Top_News/World-News/2012/05/18/North-Korea-denies-jamming-GPS-signals/UPI-65891337368638.

25. Islamic Republic of Iran Broadcasting, "Iran Hacked GPS, Hunted U.S. Drone," news release via Iran Government News (Plus Media Solutions), December 20, 2012, LexisNexis Academic.

26. David A. Fulghum, "Iranian Account of RQ-170 Loss Challenges Physics," *Aerospace Daily & Defense Report*, December 19, 2011, LexisNexis Academic.

27. Geoffrey Ingersoll, "The U.S. Deliberately Crashed a Predator Drone into an Afghanistan Mountain," blog at BusinessInsider.com, September 8, 2012, http://www.businessinsider.com/drone-piloted-by-ohioans -deliberately-crashed-in-afghanistan-2012-9.

28. Ingersoll, "U.S. Deliberately Crashed a Predator Drone."

29. Melissa Mixon, "Todd Humphreys' Research Team Demonstrates First Successful GPS Spoofing of UAV," University of Texas at Austin, news archive, June 2012, http://www.ae.utexas.edu/news/archive/2012/ todd-humphreys-research-team-demonstrates-first-successful-gps -spoofing-of-uav (accessed September 21, 2012); Michael Holden, "GPS Attacks Risk Maritime Disaster, Trading Chaos," Reuters, February 21, 2012, http://www.reuters.com/article/2012/02/22/security-gps-idUSL 5E8DK7G720120222.

30. Mixon, "Todd Humphreys' Research Team Demonstrates First Successful GPS Spoofing of UAV"; Bill Sweetman, "Student Spoofing of GPS Raises Questions," *Aerospace Daily & Defense Report*, July 3, 2012, LexisNexis Academic.

31. Holden, "GPS Attacks Risk Maritime Disaster."

32. Royal Academy of Engineering, *Global Navigation Space Systems: Reliance and Vulnerabilities* (London, March 2011), 14, http://www.raeng .org.uk/gnss.

33. Royal Academy of Engineering, *Global Navigation Space Systems*.

34. Trudy E. Bell and Tony Phillips, "A Super Solar Flare," NASA.gov, May 6, 2008, http://science.nasa.gov/science-news/science-at-nasa/2008/ 06may_carringtonflare.

35. Royal Academy of Engineering, *Global Navigation Space Systems*.

36. Dick Selwood, "What Happens When Your GPS Fails?" *Electronic Engineering Journal*, November 9, 2011, http://www.eejournal.com/ archives/articles/20111109-gps/?printView=true.

37. Bell and Phillips, "Super Solar Flare"; Royal Academy of Engineering, *Global Navigation Space Systems*.

38. U.S. Department of Defense and National Intelligence Agency, *National Security Space Strategy* (Washington DC, January 2011), 1, http://www.defense.gov/home/features/2011/0111_nsss/docs/ NationalSecuritySpaceStrategyUnclassifiedSummary_Jan2011.pdf.

39. U.S. Department of Defense and National Intelligence Agency, *National Security Space Strategy*, 1.

40. Dan Elliott, "New U.S. Satellite to Monitor Debris in Earth Orbit," Associated Press, July 3, 2010, LexisNexis Academic.

41. Naval Research Laboratory, "NRL Scientists Propose Mitigation Concept of LEO Debris," news release, June 20, 2012, http://www.nrl.navy .mil/media/news-releases/2012/nrl-scientists-propose-mitigation -concept-of-leo-debris.

42. U.S. Department of Defense and National Intelligence Agency, *National Security Space Strategy*, 2; Brian Weeden, "2009 Iridium-Cosmos Collision Fact Sheet," Secure World Foundation website, http://swfound.org/media/6575/2009%20iridium-cosmos%20 collision%20factsheet.pdf (updated November 23, 2010).

43. Weeden, "2009 Iridium-Cosmos Collision Fact Sheet"; Elliott, "New U.S. Satellite to Monitor Debris."

44. Weeden, "2009 Iridium-Cosmos Collision Fact Sheet."

45. Lisa Daniel, "Defense, State Agree to Pursue Conduct Code for Outer Space," American Forces Press Service, January 18, 2012, http://www .defense.gov/news/newsarticle.aspx?id=66833.

46. Daniel, "Defense, State Agree to Pursue Conduct Code."

47. James W. Canan, "Stepping Up Space Surveillance," *Aerospace America*, April 2007, 33, LexisNexis Academic; Fred Stakelbeck, "Red Skies," MonstersAndCritics.com, January 3, 2007, http://news.monstersand critics.com/asiapacific/features/article_1239273.php/Red_Skies.

48. Roftiel Constantine, *GPS and Galileo: Friendly Foes?* (Maxwell AFB AL: Air University Press, May 2008), 41.

49. U.S. Department of Defense and National Intelligence Agency, *National Security Space Strategy*, 2.

50. William Choong, "A Pearl Harbor in Space," *Straits Times* (Singapore), March 24, 2008, LexisNexis Academic.

51. Robert Burns, "Pentagon Concludes Navy Missile Hit Satellite's Fuel Tank and Destroyed Toxic Chemicals," Associated Press, February 25, 2008, LexisNexis Academic.

52. Walter Holemans, "Hydrazine Case Not So Hot," letter to the editor, *Aviation Week & Space Technology*, April 28, 2008, 8, LexisNexis Academic; Dave Ahearn, "Critics Assail U.S. Satellite Shoot-Down, but Stop Short of Saying It Was Unneeded," *Defense Daily*, March 25, 2008, LexisNexis Academic; Jim Mannion, "Star Wars Heating Up—U.S. Blasts Rogue Satellite," *Daily Telegraph* (Australia), February 22, 2008, LexisNexis Academic; Choong, "Pearl Harbor in Space."

53. U.S. Department of Defense and National Intelligence Agency, *National Security Space Strategy*, 2.

54. "China May Have Launched Prototype of a Maneuvering Space-Based Anti-Satellite Weapon System: Report," *Space & Missile Defense Report*, December 22, 2008, LexisNexis Academic.

55. *Encyclopedia Astronautica*, s.v. "IS-A," http://www.astronautix.com/craft/isa.htm (accessed September 20, 2012).

56. RussianSpaceWeb.com, "Spacecraft: Military: IS Anti-Satellite System," http://www.russianspaceweb.com/is.html (accessed September 20, 2012).

57. *Encyclopedia Astronautica*, s.v. "SAINT," http://www.astronautix.com/craft/saint.htm (accessed September 20, 2012).

58. Brian Weeden, "China's BX-1 Microsatellite: A Litmus Test for Space Weaponization," Space Review, October 20, 2008, http://www.thespacereview.com/article/1235/1.

59. Weeden, "China's BX-1 Microsatellite."

60. U.S. Department of Defense, *Annual Report to Congress: Military and Security Developments Involving the People's Republic of China* (Washington DC: Office of the Secretary of Defense, May 2012), www.defense.gov/pubs/pdfs/2011_cmpr_final.pdf.

61. John J. Klein, *Space Warfare: Strategy, Principles and Policy* (New York: Routledge, 2006), 68, 77.

62. Klein, *Space Warfare*, 95.

63. Klein, *Space Warfare*, 68.

64. Rajat Pandit, "After Agni-V Launch, DRDO's New Target Is Anti-Satellite Weapons," *Times of India*, April 21, 2012, http://articles.timesofindia.indiatimes.com/2012-04-21/india/31378237_1_asat-anti-satellite-agni-v; Global Security Newswire, "India Could Pursue Capability to Destroy Satellites," April 23, 2012, http://www.nti.org/gsn/article/india-wants-capability-destroy-satellites.

65. Pandit, "After Agni-V Launch, DRDO's New Target Is Anti-Satellite Weapons."

66. Ali Akbar Dareini, "Iran Builds New Space Center to Launch Satellites," Associated Press, June 2, 2012, http://bigstory.ap.org/article/iran-builds-new-space-center-launch-satellites.

67. Yiftah Shapir, "Iran's Efforts to Conquer Space," *Strategic Assessment* 8, no. 3 (November 2005), http://www.inss.org.il/publications.php?cat=21&incat=&read=160.

68. Space.com, "Iran to Launch Satellite into Space Wednesday," May 22, 2012, http://www.space.com/15824-iran-launching-fajr-satellite.html.

69. Spaceflight101.com, "Iran Suspected of Suffering Launch Failure in February," March 20, 2013, http://www.spaceflight101.com/iran-launch-failure-february-2013.html.

70. Robert Christy, "Iran's Fajr 1 Delayed . . . Again," SpaceDaily.com, May 30, 2012, http://www.spacedaily.com/reports/Iran_Fajr_1_Delayed_Again_999.html.

71. Choe Sang-Hun, "North Korean Missile Said to Have Military Purpose," *New York Times*, December 23, 2012, http://www.nytimes.com/2012/12/24/world/asia/north-korean-rocket-had-military-purpose-seoul-says.html?_r=0.

72. Weeden, "China's BX-1 Microsatellite."

73. James Clay Moltz, *The Politics of Space Security: Strategic Restraint and the Pursuits of National Interest* (Stanford CA: Stanford University Press, 2008), 336–48.

74. Gabe Starosta, "Service Declares IOC on Space Based Space Surveillance Satellite," *Inside the Air Force*, August 24, 2012, LexisNexis Academic.

75. Starosta, "Service Declares IOC on Space Based Space Surveillance Satellite."

76. Elliott, "New U.S. Satellite to Monitor Debris."

77. Starosta, "Service Declares IOC on Space Based Space Surveillance Satellite."

78. Walter Pincus, "Hearings Show Our Dependence on Military Space Technology," *Washington Post*, March 27, 2012, LexisNexis Academic.

79. U.S. General Accountability Office, *DOD Faces Challenges*, 9.

80. U.S. General Accountability Office, *DOD Faces Challenges*, 4.

81. Callan James, "What Will Follow GPS?" *Avionics*, September 2012, 22–27, http://accessintelligence.imirus.com/Mpowered/book/vav12/i9/p1.

82. DARPA.mil, "Micro-Technology for Positioning, Navigation and Timing (Micro-PNT)," http://www.darpa.mil/Our_Work/mto/Programs/Micro-Technology_for_Positioning,_Navigation_and_Timing_%28 Micro-PNT%29.aspx (accessed September 22, 2012).

83. Desiree Craig, "Truth on the Range: USAF's New Reference System," *Inside GNSS*, May–June 2012, http://www.insidegnss.com/node/3071; Locata Corporation, "About," http://www.locatacorp.com/about (accessed September 22, 2012).

84. Clay Dillow, "Ground-Based Version of Satellite GPS Could Make Positioning Technology Accurate to Inches Anywhere," *Popular Science*, August 2, 2011, http://www.popsci.com/technology/article/2011-08/ground-based-analog-gps-could-make-positioning-technology-accurate-inches#; Christopher Jay, "U.S. Approval Set to Put Locata in Top Spot," *Australian Financial Review*, July 16, 2012, LexisNexis Academic.

85. Craig, "Truth on the Range."

86. Dillow, "Ground-Based Version of Satellite GPS."

87. Marketwire, "GPS Test Director Leaves Department of Defense to Join Locata—Will Manage Locata 'Backup to GPS' Solution for Military," news release, September 12, 2012, http://www.marketwire.com/press-release/gps-test-director-leaves-department-defense-join-locata-will-manage-locata-backup-gps-1700383.htm.

88. Peter Samuel, "Swiss the First with GPS Tolling," blog at TollRoadNews.com, August 27, 2003, http://www.tollroadsnews.com/node/346.

89. Peter Samuel, "GPS-Based Toll Collect in Germany Is Collecting," blog at TollRoadNews.com, January 2, 2005, http://www.tollroadsnews.com/node/963; Peter Samuel, "Pure Satellite Tolling on Way Out in Europe—GPS/DSRC/Cellular Hybrid Coming," blog at TollRoadNews.com, January 19, 2012, http://www.tollroadsnews.com/node/5703.

90. Congressional Budget Office, *Alternative Approaches to Funding Highways* (Washington DC, March 2011), ix, http://www.cbo.gov/sites/default/files/cbofiles/ftpdocs/121xx/doc12101/03-23-highwayfunding.pdf.

91. John W. Pope Civitas Institute, "Civitas Poll: Vehicle Miles Traveled Tax Unpopular," news release, February 5, 2009, http://www.nccivitas.org/2009/civitas-poll-vehicle-miles-traveled-tax-unpopular.

92. CNN.com, "Transportation Agency: Obama Will Not Pursue Mileage Tax," February 20, 2009, http://www.cnn.com/2009/POLITICS/

02/20/driving.tax; Eric M. Weiss, "White House Dismisses Controversial Idea to Fund Transportation Projects," *Washington Post*, February 21, 2009, http://www.washingtonpost.com/wp-dyn/content/article/2009/02/20/AR2009022003331.html.

93. Keith Laing, "GOP Platform: Cut Amtrak, Privatize Airport Security and Focus Highway Money on Roads," blog at *The Hill*, August 28, 2012, http://thehill.com/blogs/transportation-report/infrastructure/246205-gop-platform-cut-amtrak-privatize-airport-security-and-focus-highway-money-on-roads.

94. Live View GPS Inc., "Proposed Vehicle Miles Traveled Tax Law in California, Using GPS Tracking," blog, July 24, 2012, http://www.liveviewgps.com/blog/proposed-vehicle-miles-traveled-tax-law-california-gps-tracking; Live View GPS Inc., "Florida Motorists May Soon Pay Miles Driven Taxes Based on GPS Tracking," blog, July 3, 2012, http://www.liveviewgps.com/blog/florida-motorists-pay-miles-driven-taxes-based-gps-tracking.

95. Larry Copeland and Paul Overberg, "States Explore New Ways to Tax Motorists for Road Repair," *USA Today*, http://www.usatoday.com/news/nation/story/2012-06-03/states-motorist-taxes/55367022/1 (updated June 5, 2012).

96. Jennifer Matarese, "Schumer Wants GPS for Truck Drivers," WABC-TV (New York), September 24, 2012, http://abclocal.go.com/wabc/story?section=news/local/long_island&id=8823069.

97. Matarese, "Schumer Wants GPS for Truck Drivers."

98. Rob Lovitt, "Connected Cars: Knight Rider Meets George Jetson," MSNBC.com, August 1, 2011, http://overheadbin.nbcnews.com/_news/2011/08/01/7199291-connected-cars-knight-rider-meets-george-jetson?lite.

99. Highway Data Loss Institute, "Crash Avoidance Features Reduce Crashes, Insurance Claim Study Shows; Autonomous Braking and Adaptive Headlights Yield Biggest Benefits," news release, July 3, 2012, http://www.iihs.org/news/rss/pr070312.html.

100. National Highway Traffic Safety Administration, "DOT Launches Largest-Ever Road Test of Connected Vehicle Crash Avoidance Technology," news release, August 21, 2012, http://www.nhtsa.gov/About+NHTSA/Press+Releases/2012/DOT+Launches+LargestEver+Road+Test+of+Connected+Vehicle+Crash+Avoidance+Technology.

101. U.S. Department of Commerce, Bureau of the Census, "Table 1103. Motor Vehicle Accidents—Number and Deaths: 1990 to 2009," in *Statistical Abstract of the United States, 2012*, http://www.census.gov/compendia/statab/2012/tables/12s1103.pdf (accessed September 2012).

102. National Highway Traffic Safety Administration, "DOT Launches Largest-Ever Road Test."

103. National Highway Traffic Safety Administration, "DOT Launches Largest-Ever Road Test."

104. Jennifer Waters, "Talking Cars Could Save Your Life," *MarketWatch*, August 9, 2012, http://www.marketwatch.com/Story/story/print?guid=AED9818C-E17C-11E1-961B-002128049AD6.

105. Waters, "Talking Cars Could Save Your Life."

106. Liviu Iftode et al., "Active Highways," paper presented at the IEEE Nineteenth International Symposium on Personal, Indoor and Mobile Communications (PIMRC), Cannes, France, September 15, 2008, http://www.google.com/url?sa=t&rct=j&q=&esrc=s&source=web&cd=1&cad=rja&ved=0CCIQFjAA&url=http%3A%2F%2Fwww.cs.rutgers.edu%2F~iftode%2Factivehw08.pdf&ei=94NgUJD5IIffogG9lICIAQ&usg=AFQjCNEnxJF_viBVmO6hm45aEZRkc29FIw.

107. Chris Taylor, "Google's Driverless Car Is Now Safer than the Average Driver," Mashable.com, August 7, 2012, http://mashable.com/2012/08/07/google-driverless-cars-safer-than-you.

108. Ben Johnson et al., "Nevada DMV Issues First License for Google's Driverless Car," blog at Slate.com, May 8, 2012, http://www.slate.com/blogs/trending/2012/05/08/google_s_driverless_car_in_nevada_gets_license_from_dmv_.html; Zusha Elinson, "Google Car Zooms Toward Legal Status," *Bay Citizen* (San Francisco Bay Area), September 24, 2012, http://www.baycitizen.org/transportation/story/google-car-zooms-toward-legal-status; Claire Cain Miller, "With a Push from Google, California Legalizes Driverless Cars," *Bits* (blog), *New York Times*, September 25, 2012, http://bits.blogs.nytimes.com/2012/09/25/with-a-push-from-google-california-legalizes-driverless-cars.

109. David Siders, "Jerry Brown Signs Driverless Car Bill," blog at *Sacramento Bee*, September 25, 2012, http://blogs.sacbee.com/capitolalertlatest/2012/09/jerry-brown-signs-driverless-car-bill.html.

110. Theo Valich, "Exclusive: Google to Certify, Sell Driverless Car Kit to the Automotive Industry," Bright Side of the News, August 15, 2012,

http://www.brightsideofnews.com/news/2012/8/15/exclusive-google
-to-certify2c-sell-driverless-car-kit-to-the-automotive-industry.aspx.

111. Alexander Besant, "Driverless Cars by Mid-century Our Likely Future, Says Technology Group," GlobalPost.com, September 23, 2012, http://www.globalpost.com/dispatch/news/science/120923/driverless-cars-mid-century-our-likely-future-says-technology-associati.

112. Dave Masko, "GPS Takes Enjoyment Out of Summertime Travel, Say Users," blog at Huliq.com, August 8, 2009, http://www.huliq.com/10282/gps-takes-enjoyment-out-summertime-travel-say-users.

113. Gilly Leshed et al., "In-Car Navigation: Engagement with and Disengagement from the Environment," paper presented at the Twenty-Sixth Computer Human Interaction (CHI) Conference, Florence, Italy, April 5–10, 2008, http://leshed.comm.cornell.edu/pubs/chi1103-leshed.pdf.

114. Véronique D. Bohbot et al., "Gray Matter Differences Correlate with Spontaneous Strategies in a Human Virtual Navigation Task," *Journal of Neuroscience* 27, no. 38 (September 19, 2007) 10078–83, http://www.jneurosci.org/content/27/38/10078.full.pdf+html.

115. Roger Highfield, "Mapping Memories: Eleanor Maguire and Brain Imaging," Wellcome Trust, http://www.wellcome.ac.uk/About-us/75th-anniversary/WTVM052023.htm (accessed September 24, 2012).

116. Highfield, "Mapping Memories."

117. Bohbot et al., "Gray Matter Differences Correlate with Spontaneous Strategies."

118. Véronique Bohbot, e-mail to Eric Frazier, September 27, 2012.

119. Bohbot, e-mail to Frazier.

120. David Goldman, "Google Unveils 'Project Glass' Virtual-Reality Glasses," CNN Money, April 4, 2012, http://money.cnn.com/2012/04/04/technology/google-project-glass/?cnn=yes&hpt=hp_mid.

121. YouTube.com, "Project Glass: One Day . . . ," April 4, 2012, http://www.youtube.com/watch?v=9c6W4CCU9M4.

122. Google.com, "Project Glass," blog post at Google+, September 13, 2012, https://plus.google.com/u/0/+projectglass/posts.

123. Russell Holly, "Google Patent Points to Smart Glove Companion for Project Glass," Geek.com, September 10, 2012, http://www.geek.com/articles/gadgets/google-patent-points-to-smart-glove-companion-for-project-glass-20120910.

124. Todd Wasserman, "Google Glass Coming to Consumer in 2014," blog

at Mashable.com, June 28, 2012, http://mashable.com/2012/06/28/brin-google-glass-2014.

125. Liat Clark, "Compared: Augmented Eyewear from Google, Apple, Olympus and Valve," Wired.co.uk, July 30, 2012, http://www.wired.co.uk/news/archive/2012-07/30/computing-glasses-comparison?page=all; Vuzix Corporation, "Vuzix Announces Advanced STAR 1200 XL See-Through Augmented Reality Glasses," news release, September 18, 2012, http://www.vuzix.com/site/_news/2012/Vuzix-Announces-STAR-1200XL.pdf; Charles Arthur, "UK Company's 'Augmented Reality' Glasses Could Be Better Than Google's," *Guardian*, September 10, 2012, http://www.guardian.co.uk/technology/2012/sep/10/augmented-reality-glasses-google-project?newsfeed=true.

126. Innovega Inc., "Innovega Demonstrates Megapixel Contact Lens Eyewear," news release, January 8, 2012, http://innovega-inc.com/press-2012.php.

127. Phys.org, "DARPA Sets Sights on High-Tech Contact Lenses," April 15, 2012, http://phys.org/news/2012-04-darpa-sights-high-tech-contact-lenses.html.

128. Daniel H. Wilson, "Bionic Brains and Beyond," *Walls Street Journal*, June 1, 2012, http://online.wsj.com/article/sb10001424052702303640104577436601227923924.html.

129. YouTube.com, *H+: The Digital Series*, http://www.youtube.com/HplusDigitalSeries?feature=etp-gs-hpl (accessed September 22, 2012).

130. Jon Orlin, "It's 2012 Already So Where Are All the Jetsons Flying Cars," TechCrunch.com, January 1, 2012, http://techcrunch.com/2012/01/01/its-2012-already-so-where-are-all-the-jetsons-flying-cars; Jack Broom, "In 1962, Third-Graders Predicted Pocket Phones, Flying Cars," *Seattle Times*, April 14, 2012, http://seattletimes.com/html/localnews/2017984268_thirdgraderlist15m.html.

Selected Bibliography

Aczel, Amir. *The Riddle of the Compass*. New York: Harcourt, 2001.

Alexander, Caroline. *The Endurance: Shackleton's Legendary Antarctic Expedition*. New York: Alfred A. Knopf, 1999.

Biskupic, Joan. "Court: GPS Tracking Needs a Warrant." *USA Today*, January 24, 2012. http://www.usatoday.com/USCP/PNI/Nation/World/2012-01-24-bcUSATSCOTUSGPSUPDT_ST_U.htm.

Bohbot, Véronique D., Jason Lerch, Brook Thorndycraft, Giuseppe Iaria, and Alex P. Zijdenbos. "Gray Matter Differences Correlate with Spontaneous Strategies in a Human Virtual Navigation Task." *Journal of Neuroscience* 27, no. 38 (September 19, 2007): 10078–83. http://www.jneurosci.org/content/27/38/10078.full.pdf+html.

Clements, William P., Jr. (deputy secretary of defense). "Memorandum for the Secretaries of the Military Departments; Subject: Defense Navigation Satellite Development Program (DNSDP)." April 17, 1973.

Coel, Margaret. "Keeping Time by the Atom." *Invention & Technology* (Winter 1998): 43–48.

Cohen, Noam. "It's Tracking Your Every Move and You May Not Even Know." *New York Times*, March 26, 2011. http://www.nytimes.com/2011/03/26/business/media/26privacy.html.

Denny, Mark. *The Science of Navigation: From Dead Reckoning to GPS*. Baltimore MD: Johns Hopkins University Press, 2012.

Divis, Dee Ann. "GPS Community Confronts LightSquared Move into L1 Spectrum." *Inside GNSS*, March–April 2011. http://www.insidegnss.com/node/2508.

———. "PNT Advisory Board Seeks Details on Economic Benefits of GPS." *Inside GNSS*, August 20, 2012. http://insidegnss.com/node/3170.

Easton, R. L. *Space Applications Branch Technical Memorandum No. 1: An Exploratory Development Program in Passive Satellite Navigation*. Washington DC: Naval Research Laboratory, May 8, 1967.

Gilroy, Amy. "Car Navigation Market Gaining Speed." *TWICE*, January 8, 2004. http://www.twice.com/article/254755-Car_Navigation_Market_Gaining_Speed.php.

GPS.gov. "National PNT Advisory Board Recommendations: White Paper on GPS Jamming." November 4, 2010. http://www.gps.gov/governance/advisory/recommendations/2010-11-jammingwhitepaper.pdf.

Hallion, Richard. *Storm over Iraq: Air Power and the Gulf War*. Washington DC: Smithsonian Institution Press, 1992.

Harvey, Brian. *The Rebirth of the Russian Space Program*. Chichester UK: Praxis Publishing, 2007.

Holmes, David C. "NAVSTAR Global Positioning System: Navigation for the Future." *Proceedings* (U.S. Naval Institute), April 1977, 101.

———. "NAVSTAR Technology." *Countermeasures* 2, no. 12 (December 1976): 53–54.

Johnson, Dana J. *Overcoming Challenges to Transformational Space Programs: The Global Positioning System (GPS)*. Arlington VA: Northrop Grumman Analysis Center, 2006.

Klass, Phillip J. "Compromise Reached on Navsat." *Aviation Week & Space Technology*, November 26, 1973, 46.

Klein, John J. *Space Warfare: Strategy, Principles and Policy*. New York: Routledge, 2006.

Lendino, Jamie. "Apple Moves to Kill GPS Devices, Reduce Dependence on Google." *PC Magazine*, June 12, 2012. http://www.pcmag.com/article2/0,2817,2405680,00.asp.

Lombardi, Michael A., Thomas P. Heavner, and Steven R. Jefferts. "NIST Primary Frequency Standards and the Realization of the SI Second." *Journal of Measurement Science* 2, no. 4 (December 2007): 74–89.

Marolda, Edward J., and Robert John Schneller Jr. *Shield and Sword: The United State Navy and the Persian Gulf War*. Annapolis MD: Naval Institute Press, 2001.

Marquis, Willard, and Michael Shaw. "GPS III: Bringing New Capabilities to the Global Community." *Inside GNSS*, September–October 2011. http://www.insidegnss.com/node/2746.

McDonald, Keith M. "Global Positioning System." In *Success Stories in Satellite Systems*, edited by D. K. Sachdev, 239–83. Reston VA: American Institute of Aeronautics and Astronautics, 2009.

Moltz, James Clay. *The Politics of Space Security: Strategic Restraint and the Pursuits of National Interest*. Stanford CA: Stanford University Press, 2008.

Naval Research Laboratory. *Timation Development Plan*. NRL Report 7227, rev. ed. Washington DC, March 2, 1971.

Nielson, John T. "The Untold Story of the CALCM: The Secret GPS Weapon Used in the Gulf War." *GPS World* 6, no. 1 (January 1995): 26–32.

O'Grady, Scott. *Return with Honor*. New York: Doubleday, 1995.

Pace, Scott, Gerald Frost, Irving Lachow, David Frelinger, Donna Fossum, Donald K. Wassem, and Monica Pinto. *The Global Positioning System: Assessing National Policies*. Santa Monica CA: Rand Corporation, 1995.

Palenchar, Joseph. "iSuppli: Smartphones Impact Seen on In-Vehicle Systems." *TWICE*, September 13, 2010. http://www.twice.com/article/457033-iSuppli_Smartphone_Impact_Seen_On_In_Vehicle_Systems.php.

Parkinson, Bradford W., and Stephen T. Powers. "Fighting to Survive: Five Challenges, One Key Technology, the Political Battlefield—and a GPS Mafia." *GPS World*, June 2010, 8–18.

Parkinson, Bradford W., and Stephen W. Gilbert. "NAVSTAR: Global Positioning System—Ten Years Later." *Proceedings of the IEEE* 71, no. 10 (October 1983): 1177.

Pham, Nam D. *The Economic Benefits of Commercial GPS Use in the U.S. and the Costs of Potential Disruption*. Washington DC: NDP Consulting Group, June 2011. http://www.saveourgps.org/pdf/GPS-Report-June-22-2011.pdf.

Rhodes, Richard. *Hedy's Folly: The Life and Breakthrough Inventions of Hedy Lamarr, the Most Beautiful Woman in the World*. New York: Doubleday, 2011.

Rip, Michael Russell, and James M. Hasik. *The Precision Revolution: GPS and the Future of Aerial Warfare*. Annapolis MD: Naval Institute Press, 2002.

Royal Academy of Engineering. *Global Navigation Space Systems: Reliance and Vulnerabilities*. London, March 2011. http://www.raeng.org.uk/gnss.

Schlesinger, James R., Laurence J. Adams, et al. *The Global Positioning System: Charting the Future, Summary Report*. Washington DC: National Academy of Public Administration and National Research Council, May 1995. http://www.navcen.uscg.gov/?pageName=gpsFuture.

Sobel, Dava. *Longitude: The True Story of a Lone Genius Who Solved the Greatest Scientific Problem of His Time*. New York: Walker Books, 1995.

"Statement by Deputy Press Secretary Larry M. Speakes on the Soviet Attack on a Korean Civilian Airliner." September 16, 1983. Public Papers of President Ronald W. Reagan, Ronald Reagan Presidential Library, http://www.reagan.utexas.edu/archives/speeches/1983/91683c.htm.

Sturdevant, Rick W. "NAVSTAR, the Global Positioning System: A Sampling of its Military, Civil, and Commercial Impact." In *Societal Impact of Spaceflight*, edited by Stephen J. Dick and Roger D. Launius, 331–52. Washington DC: NASA, 2007.

———. "The Socioeconomic Impact of the NAVSTAR Global Positioning System, 1989–2009." Paper presented at the American Institute of Aeronautics and Astronautics (AIAA) Space 2009 Conference and Exposition, September 14–17, 2009, Pasadena CA.

Taylor, Thomas. *Lightning in the Storm: The 101st Air Assault Division in the Gulf War*. New York: Hippocrene Books, 1994.

Tirpak, John A. "The Secret Squirrels." *Air Force Magazine* 77, no. 5 (April 1994): 56–60.

White House. "Statement by the President Regarding the United States' Decision to Stop Degrading Global Positioning System Accuracy." Press release, May 1, 2001. http://clinton3.nara.gov/WH/EOP/OSTP/html/0053_2.html.

Whitlock, Robert R., and Thomas B. McCaskill. *NRL GPS Bibliography: An Annotated Bibliography of the Origin and Development of the Global Positioning System at the Naval Research Laboratory*. Washington DC: Naval Research Laboratory, June 3, 2009.

U.S. Coast Guard Navigation Center. "GPS Fully Operational Statement of 1995." Press release, July 17, 1995, revised May 2, 2001. http://www.navcen.uscg.gov/?pageName=global.

U.S. Department of Defense. *Conduct of the Persian Gulf War: Final Report to Congress*. Washington DC: U.S. Government Printing Office, 1992.

U.S. Department of Defense and National Intelligence Agency. *National Security Space Strategy*. Washington DC, January 2011. http://www.defense.gov/home/features/2011/0111_nsss/docs/NationalSecuritySpaceStrategyUnclassifiedSummary_Jan2011.pdf.

U.S. General Accountability Office. *Global Positioning System: Significant Challenges in Sustaining and Upgrading Widely Used Capabilities.* GAO-09-670T. Washington DC: GAO, May 7, 2009. http://www.gao.gov/assets/130/122502.pdf.

———. *Joint Major System Acquisition by the Military Services: An Elusive Strategy.* GAO/NSIAD-84-22. Washington DC: GAO, December 23, 1983.

Werrell, Kenneth P. *Chasing the Silver Bullet: U.S. Air Force Weapons Development from Vietnam to Desert Storm.* Washington: Smithsonian Books, 2003.

Whitcomb, Darrell D. *Combat Search and Rescue in Desert Storm.* Maxwell AFB AL: Air University Press, 2006.

Woodford, J. B., and H. Nakamura. *Briefing-Navigation Satellite Study.* Los Angeles: Air Systems Command, August 24, 1966.

Index

ankle bracelets, 162–63
AN/PSN-8 "Manpack" receiver, 121–22, *123*
Antheil, George, 70
antisatellite. *See* ASAT (antisatellite)
Apollo program, 32, 33
Apple, 143–44, 147, 165, 170
armed forces, 13–14, 58–59, 63, 159–60. *See also* Air Force; Army; Navy
arms race, 199
Army, 13–14, 17, 19–20, 55
Army Map Service, 16
ARPA (Advanced Research Projects Agency), 47–48
ASAT (antisatellite), 196–98
astrolabes, 35
Atlantic, 36
atomic clocks, 28, 50, 62, 70–73, 83–87, 174. *See also* clocks
Atta, Mohammed, 151–52
Attenti, 163
augmented reality, 169, 170
automatic vehicle location. *See* AVL (automatic vehicle location)
automobiles, 1–2, 141–42, 145, 158, 202–4
autonomous cars, 203–4
aviation, 38–39, 89–92, 94, 95, 133–36, 153
AVL (automatic vehicle location), 160

backup GPS, 195
Bacon, Francis, 127
Barnes, Gene, 11
Barton, Joe, 165
Beard, Ron, 61, 65, *73*
Beidou satellites, 179–80
Berkner, Lloyd V., 7

BI (Behavior Interventions), 163
Bipartisan Congressional Privacy Caucus, 165
Blair, Charles, 38
Block II satellites, 102–3, *104*, *105*, *106*
Blue Force Tracking, 119–20
Bohbot, Véronique, 204–5
bombing, precision, 95
Bracker, Milton, 11
Bradley, Perry, 139
brain research, 204–5
Brin, Sergey, 203
Brown, Jerry, 203
brownouts, 177–78
Buckman, Dave, 176
Buisson, James, 53, 62, 65
Burke, Edmund, 41
Burrell, Gary L., 142

C/A (coarse acquisition) code, 83
CALCMS (conventional air-launched cruise missiles), 111, *112*, 113–14
Carrier IQ, 168
Carrington, Richard, 195
Carrington Event, 195–96
cars. *See* automobiles
Carter, Jimmy, 96
CBO (Congressional Budget Office), 98
CDMA (code division multiple access), 83
cell phones, 143–48, 167–68. *See also* smartphones; wristwatch phone
Cellular Telecommunications Industry Association. *See* CTIA (Cellular Telecommunications Industry Association)

Development Concept Paper Number 133, 68

DGPS (differential GPS), 132, 135, 136, 139

differential GPS. *See* DGPS (differential GPS)

Distress Alerting Satellite System, 190

Dmatek Group, 163

DNSDP (Defense Navigation Satellite Development Plan), 63–64

DNSS (Defense Navigation Satellite System), 63–73

DOD (Department of Defense), 72, 93, 94, 102, 175

Doppler measurements, 43

Doppler shift, 26

Dornan, Robert, 129

driverless cars, 203–4

Drogin, Bob, 116

drones, 152, 194

DSARC (Defense Systems Acquisition Review Council), 65, 70, 96

DuBridge, Lee, 61

Dutcher, R. L., 60

E-911 mandate, 146–48

Earhart, Amelia, 35

Easton, Roger: first satellite launch attempt and, 8; on ground station location, 62; on Lonely Halls meeting, 65; Naval Space Surveillance System and, 27, 28; with NRL team, 73; Project Vanguard and, 23–24; satellite navigation and, 45–47, 49–50, 61; on signal structure, 69–70; Timation program and, 51; on time synchronization,

54; with Vanguard TV-3 satellite, 9; Viking rocket program and, 14

EGNOS (European Geostationary Overlay System), 181

Eisenhower, Dwight D., 16–19, 22

eLoran, 200

emergency phone calls, 145–46

energy crisis, 95

Eratosthenes, 34

ESA (European Space Agency), 179, 180–81

ether theory, 44

EU (European Union), 131, 153, 154, 179, 180, 181, 196, 198

European Geostationary Overlay System. *See* EGNOS (European Geostationary Overlay System)

European Space Agency. *See* ESA (European Space Agency)

European Union. *See* EU (European Union)

Explorer I, 20, 25

F-111 fighter-bomber, 58

FAA (Federal Aviation Administration), 95, 133, 134–35, 138–39, 153, 183, 193

Faga, Martin C., 131

Falcone, Philip, 185

FBCB2 (Force XXI Battle Command Brigade and Below) program, 120

FCC (Federal Communications Commission), 82, 146, 148, 154, 158, 184, 185, 186, 193

Federal Aviation Administration. *See* FAA (Federal Aviation Administration)

Federal Communications Commission. *See* FCC (Federal Communications Commission)

federal government, public trust of, 10

Flax, Alexander H., 38–39

FOC (Full Operational Capability), declaration of, 133

Folen, Vince, 71

Force XXI Battle Command Brigade and Below program. *See* FBCB2 (Force XXI Battle Command Brigade and Below) program

Foursquare, 92

Franken, Al, 166

Franklin, Rosalind, 41–42

"freedom of space," 18–19

Full Operational Capability. *See* FOC (Full Operational Capability), declaration of

Furnas, Clifford C., 17, 18

GAGAN (GPS-Aided Geo-Augmented Navigation) system, 182

Galileo, 153–54, 179, 180–81

Galison, Peter, 71

Gallery, Daniel, 34–35

GAO (General Accounting Office), 96, 97, 98, 102, 135, 175–77, 183, 190, 200

Garmin, 142, 143

GBAS (ground-based augmentation system), 193

GDM (Generalized Development Model), 100, *101*

General Accountability Office. *See* GAO (General Accounting Office)

General Accounting Office. *See* GAO (General Accounting Office)

Generalized Development Model. *See* GDM (Generalized Development Model)

geodesy, 16

geographic information systems. *See* GIS (geographic information systems)

Geolocational Privacy and Surveillance (GPS) Act, 165–66

geophysics, 173–74

geosynchronous orbits, 43

Getting, Ivan, 61

Gibbons, Glen, 166–67

Gibbs, Asa B., 23

GIS (geographic information systems), 131, 171

Glendon, John, 57

Global Locate, 147

global navigation satellite system. *See* GNSS (global navigation satellite system)

Global Navigation Space Systems, 195

Global Positioning System. *See* GPS (Global Positioning System)

GLONASS, 173, 178–79

GNSS (global navigation satellite system), 5, 82, 155, 178, 180, 182, 183, 187, 188, 193, 195, 197, 198

Godfrey, Arthur, 64

Google, 144, 170, 203–4, 205

Google Glass, 205

Gore, Al, 138

government, public trust of, 10

GPS (Global Positioning System): civilian use of, 89–107, 130–54; commercial use of, 170–75, 201–2; development of, 2; differential,

Persian Gulf War, 109–26

personal navigation device. *See* PND (personal navigation device)

Petroski, Henry, 189

Phillips, Steven, 3

Pioneer 4, 43–44

PLGR (precision lightweight GPS receiver), 122

PND (personal navigation device), 4, 141–45, 147, 155, 158

PNT (positioning, naviation, and timing), 5, 54, 154, 176, 187, 200

Pollack, Andrew, 116

positioning, navigation, and timing. *See* PNT (positioning, naviation, and timing)

Pozesky, Martin, 135–36

PPS (Precise Positioning Service), 132, 133

Precise Positioning Service. *See* PPS (Precise Positioning Service)

precision lightweight GPS receiver. *See* PLGR (precision lightweight GPS receiver)

Presidential Decision Directive, National Science and Technology Council 6. *See* PDD NSTC-6 (Presidential Decision Directive, National Science and Technology Council 6)

Pride of Canterbury, 34

privacy issues, 159–61, 163–70

PRN (pseudorandom noise) codes, 83

Project 621B, 58, 60–61

Project Glass, 205

Project Vanguard, 7–12, 17–18, 21–29, 47

pseudorandom noise codes. *See* PRN (pseudorandom noise) codes

psychological warfare, 21–22

public relations, 175–78

public safety, 145–46

Puckett, Allen, 45

Quarles, Donald, 16, 18

quartz oscillators, 54, 62, 84, 85

Quasi-Zenith Satellite System. *See* QZSS (Quasi-Zenith Satellite System)

QZSS (Quasi-Zenith Satellite System), 182

radar, 27–28, 38, 39, 47–48, 68, 111, 118, 127, 193, 199, 203

radio waves, 38, 82–83

RAIM (Receiver Autonomous Integrity Monitoring), 134

RAND study, 136–38

range measurements, satellite navigation and, 44

READI (Real-Time Earthquake Analysis for Disaster) Mitigation Network, 174

Reagan, Ronald, 91–93

Real-Time Earthquake Analysis for Disaster Mitigation Network. *See* READI (Real-Time Earthquake Analysis for Disaster) Mitigation Network

Receiver Autonomous Integrity Monitoring. *See* RAIM (Receiver Autonomous Integrity Monitoring)

Redstone rocket program, 19–20

Reed, Frank, 37

vehicle-miles-traveled taxes. *See*
VMT (vehicle-miles-traveled)
taxes
vehicle-to-vehicle communication.
See V2V (vehicle-to-vehicle) communication
Vela Program, 97–98
video eyewear, 205–6
Viking rocket program, 13–16, 23,
46–47
"Virginia Slim" GPS receiver, 104–5
VMT (vehicle-miles-traveled) taxes,
201–2
von Braun, Wernher, 19, 20
Votaw, Martin, 8, 9, 23–24, 47

WAAS (Wide Area Augmentation
System), 133, 135–36, 139, 181, 182,
193, 195
Wales, William, 77–78
Walsh, J. Paul, 12, 20
weaponry, guided, 109–17

Weber, Joe, 71
Weeden, Brian, 199
Weems, Philip Van Horne, 38
Weiffenbach, George, 26, 27, 43
Whitlock, Robert, 219n19, 220n25,
220n33, 222n13
Wide Area Augmentation System.
See WAAS (Wide Area Augmenta-
tion System)
Wilhelm, Peter, 49, 71, 73
Williams, Don, 45
Williams, Pete, 113
Wilson, Charles H., 97
Woodford, J. B., 55, 60
Woosley, Red, 73
wristwatch phone, 161
Wyden, Ron, 165, 166

Y2K, 148–50
Y code, 83

Zakaria, Fareed, 4